现代制造技术与
食品加工装备

Modern Manufacturing Technology and
Food Processing Equipment

冯砚博　主编
郑大宇　主审

哈尔滨工业大学出版社

内 容 简 介

本书的编写目的是培养商业机械及相关专业学生解决加工装备复杂工程问题的能力。

本书共分两大部分:第一部分为现代制造技术,主要包括机械工程基本理论、计算机辅助设计与制造、建模与仿真、制造技术与装备、自动化控制技术、工业机器人、特种加工、快速成型制造、计算机集成制造、柔性制造技术、微纳制造、生物制造、先进制造理念、智能制造和制造业的发展等;第二部分为现代食品加工装备,主要包括食品加工技术和装备概论、中国传统食品加工机械、微波和超声波加工、食品工业自动控制技术及其应用、新型食品包装技术以及绿色食品加工与可持续发展等内容。

本书适用于商业机械类及其相关专业的学生使用,也可作为相关专业教师及科学和工程技术人员的参考书。

图书在版编目(CIP)数据

现代制造技术与食品加工装备/冯砚博主编
—哈尔滨:哈尔滨工业大学出版社,2020.5
ISBN 978-7-5603-8811-3

Ⅰ.①现… Ⅱ.①冯… Ⅲ.①食品加工机械-机械制造工艺 Ⅳ.①TS203

中国版本图书馆 CIP 数据核字(2020)第 081928 号

策划编辑 杨明蕾 刘 瑶
责任编辑 周一瞳 惠 晗
封面设计 刘长友
出版发行 哈尔滨工业大学出版社
社　　址 哈尔滨市南岗区复华四道街 10 号　邮编 150006
传　　真 0451-86414749
网　　址 http://hitpress.hit.edu.cn
印　　刷 哈尔滨市工大节能印刷厂
开　　本 787mm×1092mm　1/16　印张 16.25　字数 380 千字
版　　次 2020 年 5 月第 1 版　2020 年 5 月第 1 次印刷
书　　号 ISBN 978-7-5603-8811-3
定　　价 42.00 元

(如因印装质量问题影响阅读,我社负责调换)

前　言

制造业是人类社会文明的支柱产业之一。装备制造业既是基础的，又是前沿的，其高新技术层出不穷，日新月异。本书着眼于目前国内外先进、具有应用价值的技术成果，从浅入深地介绍商业机械装备所涉及的原理和技术，并基于装备制造与民生工程新认识来组织内容。

本书从基本原理到前沿技术，由浅入深地给读者一个容易接受的、全面的介绍。为加强理论的系统性和满足工程技术应用的需要，本书加入了机械设备和食品生产基础知识，温故知新，对现代先进技术内容的理解能进一步巩固和加深。为便于了解加工制造技术的发展和商业生产实际之间的密切关系，本书将其中的同类问题组织到一起，引导读者学习工程的共性技术。对这些特点进行挖掘、归纳和分析，拓宽学习思路，能充分发挥先进技术领域的融合和促进作用，完善从简单工程设备设计到复杂商用制造装备系统工程问题解决能力的知识体系，增强从整体、综合的角度开发设计有市场竞争力、适应时代需求的机械装备产品的能力。

本书分为两大部分：第一部分为现代制造技术，主要包括机械工程基本理论、计算机辅助设计与制造、建模与仿真、制造技术与装备、自动化控制技术、工业机器人、特种加工、快速成型制造、计算机集成制造、柔性制造技术、微纳制造、生物制造、先进制造理念、智能制造和制造业的发展等；第二部分为现代食品加工装备，主要包括食品加工技术和装备概论、中国传统食品加工机械、微波和超声波加工、食品工业自动控制技术及其应用、新型食品包装技术以及绿色食品加工与可持续发展等内容。

本书采用英语编写，旨在帮助读者在学习本专业英语词汇和语言表达的同时，可以提高英文科技文献的阅读能力，深化领域内的知识，为以英语为工具进行国际化交流奠定基础。书中对重要的或不熟悉的专业词汇给予了中文注解，增加了可读性。

本书由冯砚博主编，郑大宇教授主审。本书在编写过程中得到了哈尔滨商业大学机械设计教研室的杨绮云、晏祖根、金向阳、李德溥、王凤鸣、陈江、孟爽、于兴滨、孙健伟、胡广旭、纠海峰和曲云飞等同志多方面的指导和帮助，在此深表感谢。书中的内容参考了多本英文原版著作、教研室多年来培养的研究生和本科生毕业论文、教学的讲义教案及科研课题。

由于制造领域的研究发展迅速，作者对其近年发展的把握难免有疏漏，加之作者水平有限，书中难免有不妥之处，恳请广大读者批评指正。

编　者
2020 年 2 月

Contents

Chapter 0 Introduction 1
   0.1 Engineering Design 1
   0.2 Modern Manufacturing 2

Part I Modern Manufacturing Technology 7

Chapter 1 Mechanical Engineering Design 9

Chapter 2 CAD/CAM 12
   2.1 What Is CAD/CAM? 12
   2.2 Computer Aided Design 12
   2.3 Computer Aided Manufacture 14

Chapter 3 Modeling and Simulation 16
   3.1 The Role of Models in Engineering Design 16
   3.2 Mathematical Modeling 18
   3.3 Similitude and Scale Models 19
   3.4 Simulation 19
   3.5 Finite Element Analysis 20
   3.6 Computer Simulation 22
   3.7 Introduction Optimization 22

Chapter 4 Machine Tool 25
   4.1 The Lathe 25
   4.2 Drilling Machine 28
   4.3 Milling 31
   4.4 Shaping and Planing Operations 36
   4.5 Grinding 40
   4.6 Sawing Operations 42
   4.7 Broaching Operations 44
   4.8 Precision and Surface-Finishing Operations 46

Chapter 5 CNC System 50
   5.1 Introduction 50
   5.2 NC Machine Tools 50
   5.3 Computer Numerical Control 51
   5.4 CNC Machine Technical Terminologies 52
   5.5 Direct Numerical Control 52

5.6　Key Teams ………………………………………………………………… 53
Chapter 6　Automatic Control ……………………………………………………… 57
　　6.1　Open and Closed Loop Control ……………………………………………… 57
　　6.2　Fundamental Problems of Control …………………………………………… 59
　　6.3　Types of Positional Control ………………………………………………… 61
Chapter 7　Industrial Robots ……………………………………………………… 65
　　7.1　Introduction …………………………………………………………………… 65
　　7.2　Robot Configurations ………………………………………………………… 66
　　7.3　Robot Generations …………………………………………………………… 68
　　7.4　Industrial Applications of Robots …………………………………………… 68
Chapter 8　Nontraditional Machining Operations ……………………………… 71
　　8.1　Ultrasonic Machining ………………………………………………………… 71
　　8.2　Abrasive-Jet Machining ……………………………………………………… 71
　　8.3　Chemical Machining ………………………………………………………… 71
　　8.4　Electrochemical Machining ………………………………………………… 72
　　8.5　Electrodischarge Machining ………………………………………………… 72
　　8.6　Chem-Milling ………………………………………………………………… 73
　　8.7　Use of Lasers ………………………………………………………………… 73
Chapter 9　Rapid Prototyping and Manufacturing ……………………………… 76
　　9.1　Introduction …………………………………………………………………… 76
　　9.2　Rapid Prototyping and Manufacturing Technologies ……………………… 76
　　9.3　Current Application Areas of RP&M ………………………………………… 80
　　9.4　Rapid Prototyping and Manufacturing Problems …………………………… 81
Chapter 10　Computer Integrated Manufacturing ……………………………… 84
　　10.1　Introduction ………………………………………………………………… 84
　　10.2　Historical Development of CIM …………………………………………… 85
　　10.3　The CIM Wheel and Benefits ……………………………………………… 87
　　10.4　Manufacturing Resources Planning and CIM …………………………… 89
Chapter 11　Flexible Manufacturing Systems …………………………………… 91
　　11.1　Introduction ………………………………………………………………… 91
　　11.2　Flexible Manufacturing Systems …………………………………………… 91
　　11.3　Equipment …………………………………………………………………… 92
Chapter 12　Nanotechnology and Micro-Machine ……………………………… 95
　　12.1　Nanotechnology …………………………………………………………… 95
　　12.2　Micro-Machine ……………………………………………………………… 98
Chapter 13　Biofabrication ………………………………………………………… 102
　　13.1　Definition and Scope of Biofabrication …………………………………… 102
　　13.2　Practical Applications of Biofabrication Technologies …………………… 103

Chapter 14	Advanced Manufacturing Mode	106
14.1	Agile Manufacturing	106
14.2	Lean Manufacturing	108
14.3	Concurrent Engineering	110
Chapter 15	Environmentally Conscious Design and Manufacturing (ECD&M)	113
15.1	Introduction	113
15.2	Overview	113
15.3	Environmental Engineering	115
Chapter 16	Manufacturing Technology	119
16.1	Form-Changing Processes	119
16.2	Primary and Secondary Manufacturing	120
16.3	Forming	121
16.4	Separating	125
16.5	Combining	127
Chapter 17	Manufacturing Technology Today and Tomorrow	133
17.1	Introduction	133
17.2	Components of Manufacturing Systems	133
17.3	Manufacturing and Technology	135
17.4	Changes in Manufacturing	136
17.5	Timely Robots	136
Chapter 18	Intelligent Manufacturing and Case Study	138
Part Ⅱ	**Modern Food Processing Equipment**	**145**
Chapter 19	Modern Common Food Processing Technology	147
19.1	Mixing Techniques	149
19.2	Dehydration	150
19.3	Hygienic Design of Centrifugal Pumps for the Food Processing Industries	155
19.4	High Pressure Processing in Food Industry	156
19.5	Heat Transfer Mechanisms	158
19.6	Extrusion Technologies	161
19.7	Biscuit-Making Machine	164
19.8	Baking Machine	166
19.9	Conclusion	168
Chapter 20	Chinese Snack Processing	171
20.1	Noodles Processing	171
20.2	Steamed Bun Manufacture	174
20.3	Dumpling Machine	176
20.4	Hun Tun Processing	179
20.5	Sweet Dumpling Processing	180

Chapter 21	The Use of Ultrasound and Microwaves	182
21.1	Ultrasound in Food Processing	182
21.2	Ultrasound in Food Preservation	188
21.3	Ultrasound-Assisted Extraction	188
21.4	The Use of Microwaves	189
Chapter 22	Automatic Cooking Machine	193
22.1	Introduction	193
22.2	The Essential Cuisine Principle of Chinese Foods	194
22.3	The System of the Chinese Foods Cooking Machine	195
Chapter 23	Design of Robot for the Food Industry	205
23.1	Introduction	205
23.2	Design of Food Industry Robot	206
23.3	Robots in Food Industry	212
23.4	Conclusion	216
Chapter 24	Automatic Control Technology in Food Industry	219
24.1	Advanced Control Design Approaches	219
24.2	Advanced Control Strategies	220
24.3	Applications of Advanced Control Strategies in Food Processing	223
Chapter 25	Novel Food Packaging Technologies	230
25.1	Active Packaging	230
25.2	Intelligent Packaging	231
25.3	Bioactive Packaging	232
25.4	Innovative Packaging Technologies	232
25.5	Interactions of Active/Intelligent Packaging with Supply Chain	233
25.6	Nanotechnologies in Food Packaging	234
25.7	Food Safety Issues	234
25.8	Environmental Issues (Biosourced, Biodegradable, Recyclable)	235
25.9	Future Trends	235
Chapter 26	Evolution of the Food Industry—People, Tools and Machines	239
26.1	Introduction	239
26.2	Evolving Relationships between People and Their Tools and Machines	239
26.3	Machine Selection: Focus on the Product	240
26.4	Product Type: Process Details and Consumers	241
26.5	Product Contact Materials	241
26.6	Product Contact: Air and Water	242
26.7	Product Contact Surface Cleaning	243
26.8	Product Contact Surface Maintenance	243
26.9	The People Factor	244

26.10	Conclusions	244
Chapter 27	New-Generation Digitalization in Food Industry—Case Study	246
References		249

Chapter 0　Introduction

0.1　Engineering Design

To design is either to formulate a plan for the satisfaction of a specified need or to solve a specific problem. If the plan results in the creation of something having a physical reality, then the product must be functional, safe, reliable, competitive, usable, manufacturable, and marketable.

Design is an innovative and highly iterative process. It is also a decision-making process. Decisions sometimes have to be made with too little information, occasionally with just the right amount of information, or with an excess of partially contradictory information. Decisions are sometimes made tentatively, with the right reserved to adjust as more becomes known. The point is that the engineering designer has to be personally comfortable with a decision-making, problem-solving role.

Design is a communication-intensive activity in which both words and pictures are used, and written and oral forms are employed. Engineers have to communicate effectively and work with people of many disciplines. These are important skills, and an engineer's success depends on them.

A designer's personal resources of creativeness, communicative ability, and problem-solving skill are intertwined with the knowledge of technology and first principles. Engineering tools (such as mathematics, statistics, computers, graphics, and languages) are combined to produce a plan that, when carried out, produces a product that is functional, safe, reliable, competitive, usable, manufacturable, and marketable, regardless of who builds it or who uses it.

All design activities must do the following:

(1) Know the "customers' needs".

(2) Define the essential problems that must be solved to satisfy the needs.

(3) Conceptualize the solution through synthesis, which involves the task of satisfying several different functional requirements using a set of inputs such as product design parameters within given constraints.

(4) Analyze the proposed solution to establish its optimum conditions and parameter settings.

(5) Check the resulting design solution to see if it meets the original customer needs.

Other problems require far more imaginative ideas and inventions for their solution. The word "creativity" has been used to describe the human activity that results in ingenious or unpredictable or unforeseen results (e. g. , new products, processes, and systems). Design will always benefit when "creativity" or "inspiration" and/or "imagination" plays a role.

Many problems in mechanical engineering can be solved by applying practical knowledge of engineering, manufacturing and economics.

0.2 Modern Manufacturing

1. Modern Manufacturing Technology

Manufacturing is the conversion of raw materials into desired end products. The word derives from two Latin roots meaning hand and make. Manufacturing, in the broad sense, begins during the design phase when judgments are made concerning part geometry, tolerances, material choices, and so on. Manufacturing operations start with manufacturing planning activities and with the acquisition of required resources, such as process equipment and raw materials. The manufacturing function extends throughout a number of activities of design and production to the distribution of the end product and, as necessary, life cycle support.

Manufacturing technologies address the capabilities to design and to create products, and to manage that overall process. Product quality and reliability, responsiveness to customer demands, increased labor productivity, and efficient use of capital were the primary areas that leading manufacturing companies throughout the world emphasized during the past decade to respond to the challenge of global competitiveness. As a consequence of these trends, leading manufacturing organizations are flexible in management and labor practices, develop and produce virtually defect-free products quickly (supported with global customer service) in response to opportunities, and employ a smaller work force possessing multi-disciplinary skills. These companies have an optimal balance of automated and manual operations.

To meet these challenges, the manufacturing practices must be continually evaluated and strategically employed. In addition, manufacturing firms must cope with design processes (e. g. , using customers' requirements and expectations to develop engineering specifications, and then designing components), production processes (e. g. , moving materials, converting materials properties or shapes, assembling products or components, verifying processes results), and business practices (e. g. , turning a customer order into a list of required parts, cost accounting, and documentation of procedures). Information technology will play an indispensable role in supporting and enabling the complex practices of manufacturing by providing the mechanisms to facilitate and manage the complexity of manufacturing processes and achieving the integration of manufacturing activities within and among manufacturing enterprises. A skilled, educated work force is also a critical component of a state-of-the-art manufacturing capability. Training and education are essential, not just for new graduates, but for the existing work force.

Manufacturing is evolving from an art or a trade into a science. The authors believe that we must understand manufacturing as a technical discipline. Such knowledge is needed to most effectively apply capabilities, quickly incorporate new developments, and identify the best available solutions to solve problems. The structure of the science of manufacturing is very similar across product lines since the same fundamental functions are performed and the same basic managerial controls are exercised.

2. Manufacturing Systems and Processes Design

(1) Manufacturing systems design.

The design of a manufacturing system may be defined in terms of the concept of domains and the integration of three or more design fields; they are product design, organization design, and software design.

①Product design. The design of the direct manufacturing operations is based on mapping the product's design parameters in the physical domain to process variables in the process domain.

②Organizational design. The human and financial resources, which contribute to both direct operations and the indirect overhead, are provided to satisfy the goals and structure of the business organization.

③Software design. Computer-based tools are easily available to control production and inventory levels, support information flow for decision-making, and automate manufacturing operations.

(2) Manufacturing process design.

The principal objective of manufacturing process design is to produce an organized plan for converting raw materials into useful products. It involves the selection of timely and cost-effective methods to produce a product without compromising quality and reliability. As part of the product development process, good manufacturing process design contributes to the industrial competitiveness of a manufacturing enterprise, while poor process design contributes to cost and schedule overruns and the delivery of products that fail to meet some or all of the customers' needs.

3. Advanced Manufacturing

Since the beginning of the 21st century, new-generation information technology has shown explosive growth. The broad application of digital, networked, and intelligent technologies in the manufacturing industry and the continuous development of integrated manufacturing innovations have been the main driving forces of the new industrial revolution. In particular, new-generation intelligent manufacturing, which serves as the core technology of the current industrial revolution, incorporates major and profound changes in the development philosophy, manufacturing modes, and other aspects of the manufacturing industry. Advanced manufacturing technology (AMT) is important for improving manufacturing system competitiveness.

Manufacturing technologies traditional consists of five categories, namely joining, dividing, subtractive, transformative and additive technologies.

(1) Joining technology. Consists of processes by which two or more workpieces are joined to form a new workpiece.

(2) Dividing technology. Dividing processes are the opposite of joining processes.

(3) Subtractive technology. Subtractive/Negative operations are material removal processes, by which material is removed from a single workpiece, resulting in a new workpiece.

(4) Transformative technology. A single workpiece is used to create another workpiece and the mass does not change.

(5) Additive technology. Material is added to an existing workpiece to build a new workpiece, where the mass of the finished workpiece is greater than before.

Advanced manufacturing is a family of activities that: depend on the use andcoordination of information, automation, computation, software, sensing, and networking; make use of cutting-edge materials and emerging capabilities enabled by the physical and biological sciences, for example, nanotechnology, chemistry, and biology. It involves new ways to manufacture existing products, and especially the manufacture of new products emerging from new advanced technologies.

Advanced manufacturing relies on new technologies that enable flexibility, agility, and the ability to reconfigure, as for example the following applicable areas: bio-manufacturing, semiconductors, advanced materials, additive manufacturing, and nano manufacturing. Emerging technologies can have game-changing impacts on manufacturing models, approaches, concepts, and even businesses.

Industry 4.0, a German strategic initiative, is aimed at creating intelligent factories where manufacturing technologies are upgraded and transformed. Intelligent manufacturing plays an important role in Industry 4.0. Typical resources are converted into intelligent objects so that they are able to sense, act, and behave within a smart environment. It is associated with cyber-physical systems (CPSs), artificial intelligence (AI), the internet of things (IoT), cloud computing, big data analytics (BDA), and information and communications technology (ICT). Intelligent manufacturing takes advantage of advanced information and manufacturing technologies to achieve flexible, smart, and reconfigurable manufacturing processes in order to address a dynamic and global market. It enables all physical processes and information flows to be available when and where they are needed across holistic manufacturing supply chains, multiple industries. Intelligent manufacturing is now reshaping the development paths, technical systems, and industrial forms of the manufacturing industry, and is thereby pushing the global manufacturing industry into a new stage of development.

Industry 4.0 is a complex and flexible system. This challenge becomes even more important for the food industry, which is considered one of the most important sectors of the current economy. There is an increasing level of variability in terms of demand, volume, process, manufacturing technology, customer behaviour and supplier attitude. So the food segment is facing peculiar global challenges where the new paradigm of Industry 4.0 can represent an interesting evolution. Specifically, the food industry has changed from a supply-based approach to a demand-based approach, the so-called "chain reversal", in which production will be tailored to customer demand. Tastes differ and eating and drinking are getting more individual. In order to realize this vision, elements such as machines, storage systems, and utilities must be able to share information, as well as act and control each other autonomously. Various kinds of new technologies are emerging, such as digital factory, digital food enterprise, factory of the future, smart manufacturing, and intelligent packaging. Moreover, in developed economies, firms are embracing these advanced technologies following a sequential upgrading strategy—from digital manufacturing to smart manufacturing (digital-networked), and then to new-generation intelligent manufacturing paradigms.

Part Ⅰ Modern Manufacturing Technology

Chapter 1   Mechanical Engineering Design

Mechanical engineers are associated with the production and processing of energy, providing the means of production, the tools of transportation, and the techniques of automation. The skill and knowledge base are extensive. Among the disciplinary bases are mechanics of solids, fluids, mass and momentum transport, manufacturing processes, electrical and information theory. Mechanical engineering design involves all the disciplines of mechanical engineering.

Problems resist compartmentation. A simple journal bearing involves fluid flow, heat transfer, friction, energy transport, material selection, thermomechanical treatments, statistical descriptions, and so on. A building is environmentally controlled. The heating, ventilation, and air-conditioning considerations are sufficiently specialized that some speak of heating, ventilating, and air-conditioning design as if it is separate and distinct from mechanical engineering design. Similarly, internal-combustion engine design, turbo-machinery design, and jet-engine design are sometimes considered discrete entities. The leading adjectival string of words preceding the word design is merely a product-descriptive aid to the communication process. There are phrases such as machine design, machine-element design, machine-component design, systems design, and fluid power design. All of these phrases are somewhat more focused examples of mechanical engineering design. They all draw on the same bodies of knowledge, are similarly organized, all require similar skills.

In the academic world, with its clustering of knowledge into efficient learning groups, we encounter subjects, courses, disciplines, and fields. Curricula consist of sequences of courses. The arrangement of courses presents the opportunity to study machine elements and machines earlier than the last semester. Thus machine design often represents the student's first serious design experience with a substantial knowledge base. Some, but not many machine elements can be understood without a complete thermofluid base, but before you know it, we are into mechanical engineering design.

Science explains what is, engineering creates what never was. Mathematics is neither science nor engineering. Physics and chemistry are science, but not engineering. As suggested in Figure 1.1, it takes one kind of talent to be a scientist and a different talent to create what never was. Engineers and scientists know something of each other's work, but only in rare cases are both talents developed in an individual. It takes talent and ability to create and innovate, talent to be a consistently successful problem solver and decision maker, and talent to be an effective communicator. Preparation, you see, is the developing and polishing of talent, whatever the endeavor.

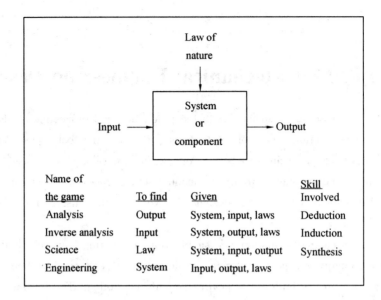

Figure 1.1 The name(s) of the game(s). Note distinctions between analysis science, and engineering and the significant skills involved

Design situations previously encountered in your curriculum drew on the very small information base available, and the idea was to briefly present the design side of engineering. The prerequisite base of this book, however, is formidable. The amount of relevant details is now sufficiently large that we must employ formalisms which organize, permit insights, and allow reduction in clutter. The individual talent and creativeness needed now are larger than before. If your talents are not developing as rapidly as you would like, explore with your instructor ways of enhancing your development and realizing your potential. There are good innovative engineers who are not great analysts, and fine analysts who cannot innovate. The world needs both, and they work together.

Viewpoint to an engineer is to be able to be in a position to have a commanding view of relevant things. This view has to be communicated to others. Viewpoint is not just an important thing; viewpoint is everything. As engineers "talk to themselves", then, to the extent that humans think with words, words and their meanings can be helpful or hurtful. If we do not have the words, or do not know the words, key thoughts often do not penetrate our stream of consciousness. Engineers also use geometric thinking and drawings because verbal language is inadequate to the task.

Since words are very important in engineering communication, we should be aware of the debasement of words, their meaning, and representations of reality brought about by our culture. Children and many adults have difficulty distinguishing between fiction and reality. Consider television commercials, which are almost entirely false. The repairman is not a repairman but a hired impersonator, espousing a provided text, crafted to persuade rather than inform, and delivered on a sham set. Language can be used to inform, or it can be used to persuade. Language used to inform places all relevant factors before the listener, so that a person is in a position to think and to bring that information to bear on a solution to a problem. Language used to persuade employs selected facts which support a bias, and omission is a standard tactic.

Designers have to make decisions, few or many, some apriority, some in concert. We need a bookkeeping system to track where we are, a useful vocabulary, a shorthand of "pegs" upon which to hang kindred ideas.

Words/Phrases and Expressions

mechanical engineering 机械工程
compartmentation 分隔
ventilation 通风
internal-combustion engine 内燃机
thermofluid 热流体
formalism 形式,形式主义
law of nature 自然法则
prerequisite 先决条件
penetrate 穿透,渗入,贯穿
consciousness 意识
geometric thinking 几何思维
impersonator 模仿者
espouse 拥护,支持
peg 钉子,楔子,橛子,短桩,夹子
shorthand 速记,简略
kindred 类似的,相似的,相关的

Chapter 2　CAD/CAM

2.1　What Is CAD/CAM?

The term CAD/CAM stands for Computer Aided Design and Computer Aided Manufacture, and is intended to relate to a closely integrated system of producing components from the design stage through to manufacture using computer assistance. In practice it is only larger installations that achieve this goal. It is more common to find that CAD systems (based on computer aided draught principles), and CAM systems (based on stand-alone CNC machines), operate in comparative isolation. True CAD/CAM systems provide the all-important link that converts information generated by the CAD system into a form that can be readily utilized by the manufacturing systems of the organization.

2.2　Computer Aided Design

CAD systems require large amounts of computing power. For this reason they are usually based around mini-computers or large-capacity microcomputers. Peripheral devices such as printers, plotters, disc storage devices, etc., are also required. These may be shared by a number of users. CAD Software is usually expensive.

It is common to talk of a CAD workstation at which the designer interacts with the system. A typical workstation comprises a high-resolution graphics screen and some means of inputting information. Input devices include key-boards, graphics/menu tablets, light pens, joysticks and hand-held "mice". On larger systems it is common to have an additional VDU screen on which commands, menus and other textual information appears, all graphics being plotted on a dedicated graphics VDU. The designer using a CAD system now spends most of the working day sitting at the computerized workstation and concentrating on information displayed on VDU devices. Many of the above input devices have been designed to ease the strain of interacting with computerized systems. These and other ergonomic considerations, such as lighting and posture, should be considered as an important aspect in the planning of a CAD installation.

CAD software is usually supplied as different packages that can be used or different aspects of the design process. A CAD system is likely to consist of more than one package. It is essential that all packages within the system are compatible so that data can be transferred between them. The most common CAD packages include followings.

1. Design Analysis Packages

This is specialist software that performs complex mathematical analyses on designs originated by the designer. Stress analysis on large structures is a typical example. Such packages have the advantage that designs can be optimized for best performance in a very short time. Better quality products with fewer errors can be produced quickly and accurately.

2. Surface and Solid Modeling Packages

This software allows the creation of three-dimensional models of objects to be constructed quickly on a graphics screen. Design concepts can be visualized from all angles, in true perspective and, in some cases, colour, surface modeling generally produces wire-frame representations, whilst solid modeling produces a solid representation which can be sectioned as required. Modifications to the design can be carried out swiftly allowing various alternatives to be explored. Customers' requirements can be matched quickly and accurately and design time is considerably shortened.

3. Simulation Packages

This software usually models the operation of dynamic systems. For example, mechanical systems such as car suspension performance, or organizational systems such as complete factory layouts simulating different component routing through the system, may be simulated. Performance characteristics and design variables can be optimized in a very short time without the need to construct physical set-ups.

4. 2-D and 3-D Draughting Packages

These are by far the most common packages in use. They allow detail drawings and designs to be constructed quickly and easily on the graphics screen. Complete working drawings can then be generated from the stored information.

In addition to the facilities of:

(1) Move, draw and delete lines;
(2) Points, circles, arcs, etc. in a variety of line types;
(3) Ability to store and retrieve information to/from backing store.

Common facilities of such systems also include the following:

(1) Rotate all or part of the design through any angle;
(2) Scale parts of the design up or down;
(3) Translate parts of the design to different positions;
(4) Replicate the same feature(s) at different places;
(5) Pan across a large design using the screen as a window;
(6) Zoom into or away from design features.

In addition, many packages offer automatic dimensioning and cross-hatching facilities and the ability to apply layering. Layering involves producing different parts of the design on different screens and storing them separately. The designer can then recall any screen combination and superimpose different layers on top of each other. This allows an uncluttered build-up of drawings to be accomplished.

Standard parts or symbols may also be defined and stored in a symbol library. They may then be recalled and positioned on the drawing at any desired position.

The data generated by the software is held on disc or tape as computer files. If this data can be accessed by other software within the manufacturing system, it can then form the basis of a manufacturing database that can support post-design activities.

2.3　Computer Aided Manufacture

The most readily understood form of CAM is the stand-alone CNC machine tool, closely followed by DNC operation. The definition of CAM, however, should be understood in a wider context. Non-metal-cutting production processes(welding, presswork, EDM, etc.), and related activities such as computer aided inspection, testing and assembly techniques, together with automated materials handling, should be included.

CAM software is perhaps just as important. Much CAM software makes use of data generated at the CAD stage. By interrogating the appropriate computer files, CAM software can provide:

(1) Bill of material and parts list generation which can be used for materials requirements planning(MRP), stock control and estimating.

(2) Post-processed output of component design data into CNC part program language suitable for transmission via DNC.

(3) Production planning data concerning machining times, operation sequences, and tooling requirements which can assist process and production planning.

Much of the above CAM software is itself now being packaged into a separate category known as Computer Aided Production Management(CAPM). The aim of a CAPM system is to monitor and control the production management elements of materials control, work-in-progress, and accounting.

It can be seen that a manufacturing installation making use of CAD/CAM. FMS and CAPM facilities become a totally integrated manufacturing system. Benefits come from each sector of the manufacturing facility working on the same, current and up-to-date information from a system database. Such information is available immediately and much of the routine and time-consuming clerical work is dealt with by computer. The majority of paperwork, documentation and records are also computer-generated.

Rarely do such systems become established by design. They are usually formed by introducing computer-based facilities into existing set-ups. The benefits derived from installing individual systems cannot be equated to the full potential of a truly integrated system. Reasons for this include: missing links in the manufacturing chain; already established work practices, procedures and systems; traditional departmental boundaries; resistance to change, etc..

The modern production engineering philosophy advocates a total Computer Integrated Manufacturing (CIM) approach as a total design concept. Such approaches are rare since the amount of preparatory work in design and planning, and the huge capital investment required in plant and machine tools, are often prohibitive.

Words/Phrases and Expressions

computer aided design(CAD) 计算机辅助设计
computer aided manufacture(CAM) 计算机辅助制造
draught 绘图,草图
stand-alone 独立的
large-capacity 大容量
microcomputer 小型计算机
high-resolution 高分辨率
graphic 图形,图形的
screen 显示器
key-board 键盘
light pen 光笔
joystick 游戏杆
hand-held 手持
mice (mouse 的复数) 鼠标
peripheral device 外围设备
workstation 工作站
interact 交互
ergonomic 人体工程学
package 软件包
compatible 兼容的
surface and solid modeling packages 曲面和实体建模包
layout 布局
non-metal-cutting produce 非金属切削生产
welding 焊接
presswork 冲压
EDM 电火花加工
Materials Requirements Planning(MRP) 物料需求计划
Computer Aided Production Management(CAPM) 计算机辅助生产管理

Chapter 3　Modeling and Simulation

3.1　The Role of Models in Engineering Design

A model is an idealization of a real-world situation that aids in the analysis of a problem. You have employed models in much of your education, and especially in the study of engineering you have learned to use and construct models such as the free-body diagram, electric circuit diagram, and the control volume in a thermodynamic system.

A model may be either descriptive or predictive. A descriptive model, enables us to understand a real-world system or phenomenon; an example is a cutaway model of an aircraft gas turbine. Such a model serves as a device for communicating ideas and information. However, it does not help us to predict the behavior of the system. A predictive model is used primarily in engineering design because it helps us to both understand and predict the performance of the system.

We also can classify models as follows:

(1) Static-dynamic;

(2) Deterministic-probabilistic;

(3) Iconic-analog-symbolic.

A static model is one whose properties do not change with time; a model in which time-varying effects are considered is dynamic. In the deterministic-probabilistic class of models there is differentiation between models that predict what will happen. A determinate mode describes the behavior of a system in which the outcome of an event occurs with certainty. In many real-world situations the outcome of an event is not known with certainty, and these must be treated with probabilistic models.

An iconic model is one that looks like the real thing. Examples are a scale model of an aircraft for wind tunnel test and an enlarged model of a polymer molecule. Iconic models are used primarily to describe the static characteristics of a system, and they are used to represent entities rather than phenomena. As geometric representations they may be two-dimensional (maps, photographs, or engineering drawings) or three-dimensional (a balsa wood and paper model airplane or a wood and plastic mockup of a full-size automobile). Three-dimensional models are especially important to communicate a complex design concept, gage customer reaction to design styling, study the human engineering aspects of the design, and check for interferences between parts or components of a large system.

We can distinguish four types of physical or iconic models that are used in engineering design. The proof of concept models a minimally operative model of the basic principle of the design concept. It is usually very elementary and assembled from readily available parts and materials. Sometimes this is known as a "string and chewing gum" model. The scale model is dimensionally shrunken or enlarged compared with the physical world. It is often a nonoperating model made from wood or plastic, but it is important for communicating the concept and for visualizing possible space interferences or conflicts. The experimental model is a functioning model containing the ideas of the design concept. It is as nearly like the proposed design as possible, but it may be incomplete in appearance. This model is subjected to extensive testing and modification. The prototype model is a full-scale working model of the design. It is technically and visually complete. The prototype often is handmade, but in other respects is intended to meet the needs of the user. As we move from proof of concept to prototype the model increases in complexity, completeness, and cost.

Analog models are those that behave like real systems. They are often used to compare something that is unfamiliar with something that is very familiar. Unlike an iconic model, an analog model need look nothing like the real system it represents. It must either obey the same physical principles as the physical system or simulate the behavior of the system. There are many known analogies between physical phenomena, but the one most commonly used is the analogy between easily made electrical measurements and other physical phenomena. An ordinary graph is really an analog model because distances represent the magnitudes of the physical quantities plotted on each axis. Since the graph describes the real functional relation that exists between those quantities, it is a model. Another common class of analog models are process flow charts.

Symbolic models are abstractions of the important quantifiable components of a physical system. A mathematical equation expressing the dependence of the system output parameter on the input parameters is a common symbolic model. A symbol is a shorthand label for a class of objects, a specific object, a state of nature, or simply a number. Symbols are useful because they are convenient, add to simplicity of explanation, and increase the generality of the situation. A symbolic model probably is the most important class of model because it provides the greatest generality in attacking a problem. The use of a symbolic model to solve a problem calls on our analytical, mathematical, and logical powers. A symbolic model is also important because it leads to quantitative results. We can further distinguish between symbolic models that are theoretical model, which are based on established and universally accepted laws of nature, and empirical models, which are the best approximate mathematical representations based on existing experimental data.

We have seen that modeling is the representation of a system or part of a system in physical or mathematical form that is suitable for demonstrating the behavior of the system. Simulation involves subjecting models to various inputs or environmental conditions to observe how they behave and thus explore the nature of the results that might be obtained from the real-world system. Simulation is manipulation of the model. It may involve system hardware (prototype models) subjected to the actual physical environment, or it may involve mathematical models subjected to mathematical disturbance functions that simulate the expected service conditions.

The solution of models by the straightforward application of mathematical techniques has been the classical approach in much academic engineering education. However, only the simplest (and hence most unrealistic) models can be solved with classic analytic methods. The widespread use of the digital computer, and to a much lesser extent the analog computer, has greatly expanded the scope and usefulness of mathematical modeling. The use of numerical methods for solution and the ease with which iterative procedures can test many specific states of the model have firmly established computer modeling and simulation as a powerful tool of engineering design.

Engineers use models for thinking, communications, prediction, control, and training. Since many engineering problems deal with complex situations, a model often is an aid to visualizing and thinking about the problem. One of the results of an engineering education is that you develop a "menu of models" that are used instinctively as part of your thought process. Models are vital for communicating whether via the printed page, the computer screen, or oral presentation. Generally, we do not really understand a problem thoroughly until we have predictive ability concerning it. Engineers must make decisions concerning alternatives. The ability to simulate the operation of a system with a mathematical model is a great advantage in providing sound information, usually at lower cost and in less time than if experimentation had been required.

Moreover, there are situations in which experimentation is impossible because of cost, safety, or time. While we usually worry about whether the model describes the real situation closely enough, there are situations in which the model constrains and controls the real-world system. For example, a detailed engineering drawing is an iconic model that describes in complete detail how the part is to be built. Finally, we have the growing use of detailed models for the training of operators of complicated systems. Thus, airline pilots train on flight simulators and nuclear power plant operators learn from reactor simulators.

3.2 Mathematical Modeling

In mathematical modeling the components of a system are represented by idealized elements that have the essential characteristics of the real components and whose behavior can be described by mathematical equations. The first step is to devise a conceptual model that represents the real system to be analyzed. You have been exposed to many examples of simple mathematical models in your engineering courses, but modeling is a highly individualized art. A key issue is the assumptions, which determine on the one hand the degree of realism of the model and on the other hand the practicality of the model for achieving a numerical solution.

Skill in modeling comes from the ability to devise simple yet meaningful models and to have sufficient breadth of knowledge and experience to know when the model may be leading to unrealistic results.

For example, in a mathematical model for a system or component, the choice of the system that is modeled is an important factor in the success of the model. Engineering systems frequently are very complex. Progress often is better made by breaking the system into simpler components and modeling each of them. In doing that, allowance must be made for the interaction of the components with each other. Techniques for treating large and complex systems by isolating the critical components and modeling them are at the heart of the growing discipline called systems engineering.

3.3 Similitude and Scale Models

New processes and designs are not born big, they grow through laboratory, development, and pilot plant stages. Physical models are an important part of the development and design processes. Usually scale models, which are less than full size, are employed. For example, in a model for aerodynamics wind tunnel testing, a pilot plant reproduces in reduced scale all or most aspects of a chemical, metallurgical, or manufacturing process.

In using physical models it is necessary to understand the conditions under which similitude prevails for both the model and the prototype. Geometric similarity is the first type of similitude. The conditions for it are a three-dimensional equivalent of a photographic enlargement or reduction, i.e., identity of shape, equality of corresponding angles or arcs, and a constant proportionality or scale factor relating corresponding linear dimensions.

$$\text{Model dimension} = \text{scale factor} \times \text{prototype dimension} \tag{3.1}$$

3.4 Simulation

Simulation involves the modeling of a complex situation into a simpler and more convenient form that can be studied in isolation without the troublesome complex side effects that usually accompany a real engineering situation. The purpose of the simulation is to explore the various outputs that might be obtained from the real system by subjecting the model to environments that are in some way representative of the situations it is desired to understand. Simulation invariably involves the use of the computer to perform often laborious computations and to follow the dynamics of the situation. The example of a process model included a simulation of the brake bending process that utilized a computer program interactive with graphics display. We shall briefly consider some additional aspects of computer simulation.

Computer simulations involve a time dimension (dynamic behavior), and they fall into three broad categories.

(1) The simulation of an engineering system or process by mathematical modeling and computer simulation. An example would be the simulation of a traffic control problem or the solidification of a large steel casting.

(2) Simulation gaming (not to be confused with game theory), in which live decision makers use a computer model for complex situations involving military or management decisions. An example would be a game for bidding strategy in the construction industry.

(3) Simulation of business or industrial engineering systems, which includes such problems as control of inventory levels, job-shop scheduling, and assembly-line balancing.

3.5 Finite Element Analysis

The finite-element method is a powerful computer-based method of analysis that is finding wide acceptance for the realistic modeling of many engineering problems. Here it is presented in some detail so you can appreciate its mathematical structure and gain some appreciation of how it is used in computer modeling.

In finite element analysis a continuum solid or fluid is considered to be built up of numerous tiny connected elements (Figure 3.1). Since the elements can be arranged in virtually any fashion, they can be used to model very complex shapes. Thus, it is no longer necessary to find an analytical solution that treats a close "idealized" model and guess at how the deviation from the model affects the prototype. As the finite-element method has developed, it has replaced a great deal of expensive preliminary cut-and-try development with quicker and cheaper computer modeling.

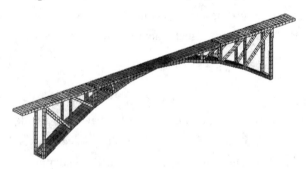

Figure 3.1 Simple finite-element representation of a beam

In contrast to the analytical methods that often require the use of higher-level mathematics, the finite-element method is based on simple algebraic equations. However, an FEM solution may require hundreds of simultaneous equations with hundreds of unknown terms. Therefore, the development of the technique required the availability of the high-speed digital computer for solving the equations efficiently by matrix methods. The rapid acceptance of finite-element analysis has been largely due to the increased availability of FEM software through interactive computer systems.

In the finite-element method, the loaded structure is modeled with amesh of separate elements. We shall use triangular elements here for simplicity, but later we shall discuss other important shape elements. The elements are connected to one another at their corners, and the connecting points are called nodes. A solution is arrived at by using basic stress and strain equations to compute the deflections in each element by the system of forces transmitted from neighboring elements through the nodal points. The strain is determined from the deflection of the nodal points, and from the strain the stress is determined with the appropriate constitutive equation. However, the problem is more complex than first seen, because the force at each node depends on the force at every other node. The elements behave like a system of springs and deflect until all forces are in equilibrium. That leads to complex system of simultaneous equations. Matrix algebra is needed to handle the cumbersome systems of equations.

The computer analysis usually is carried through to compute the principal stresses and their directions throughout the part. The printed volume of computer output for complex models is enormous and difficult to handle. A graphics display to display final output data through stress contours, color graphics, etc., is very helpful.

Finite element analysis was originally developed for two-dimensional (plane stress) situations. A three-dimensional structure causes orders of magnitude increase in the number of simultaneous equations; but by using higher-order elements and faster computers, these problems are being handled by the FEM.

A cumbersome part of the FEM solution is the preparation of the input data. The topology of the element mesh must be described in the computer program with the node numbers and the coordinates of the node points, along with the element numbers and the node numbers associated with each element. This bookkeeping task is extremely tedious for an FEM model containing hundreds of nodes. If the data must be prepared by hand and input with cards, the job is very time consuming and subject to errors. To avoid it, many large users of the FEM have developed automatic mesh generation techniques for producing the system topology. Usually, these techniques produce meshes containing only one kind of element.

The key to the practical utilization of the FEM is the finite-element model. For sound economics the model should contain the smallest number of elements to produce the needed accuracy. The best procedure is to use an iterative modeling strategy whereby coarse meshes with few elements are increasingly refined with fine meshes in critical areas of the model. Coarse models can be constructed with one-dimensional simple beams and two-dimensional rectangular (plate) elements. Rather than depict the true geometry of the part, the coarse model represents only how the structure reacts to loads. Small holes, ribs, flanges, and similar details are purposely ignored. Computer costs for finite-element analysis increase exponentially with the number of elements.

It is not necessary to develop your own computer program to use the FEM. General purpose programs are available from a number of commercial sources or can be obtained at nominal cost from the federal government. The extensive documentation that is available makes the programs relatively easy to master. However, one should realize that using a versatile general-purpose program to solve a specialized problem may be far more costly than developing a program specifically for the purpose.

3.6 Computer Simulation

The user of computers to carry out extensive simulations involving graphical output has become commonplace. In fact, many people are referring to these as computer experiments, and there is the prediction that this field will eventually grow into a third domain of science, coequal with the traditional domains of theory and experimentation. In doing these simulations finite element methods are by far the predominant method of analysis. A major impetus to the growth of this field is the availability of very high speed computers. For example, the use of a supercomputer to perform an airflow analysis on an automobile required only 20 minutes of computational time, while the same analysis using a large mainframe required 10 hours.

For example, computer simulation is used to study the deflection of the bottom dome of an aluminum can after it strikes the ground. For the simulated drop test on a supercomputer the computer is fed the shape of the can, the speed at which it hits the ground, the mass and pressure of the liquid inside the can, and the mechanical properties of the aluminum sheet. The finite element analysis combines the equations that describe the complicated interplay of mass, velocity, acceleration, pressure, and mechanical behavior to calculate the display the distortion of the can at one millisecond intervals after impact. Previous to using computer simulation a conventional development approach was used in which experimental designs were fabricated and tested. A typical program would take six months to a year and cost around $100 000, whereas with computer modeling it takes less than two weeks at a cost of about $2 000. The same kind of cost saving carries over to larger, more complex systems. A major US auto maker used supercomputer simulation to minimize noise and vibration in the 1986 models. They were able to investigate about 20 different designs for the body and suspension system in the time it would have taken to build three prototypes, at a cost saving of $3 million.

3.7 Introduction Optimization

We have continually emphasized that design is an iterative process. You start with a poorly defined problem, refine it, develop a model, and arrive at a solution. Usually there is more than one solution and the first one is not usually the best. Thus, in engineering design we have a situation in which there is a search for the best answer. In other words, optimization is inherent in the design process. A mathematical theory of optimization has been developed since 1950, and it has gradually been applied to a variety of engineering design situations. The concurrent development of the digital computer, with its inherent ability for rapid numerical calculation and search, has made the utilization of optimization procedures practical in many design situations.

By the term optimal design we mean the best of all feasible designs. Optimization is the process of maximizing a desired quantity or minimizing an undesired one. Optimization theory is the body of mathematics that deals with the properties of maxima and minima and how to find maxima and minima numerically. In the typical design optimization situation, the designer has created a general configuration for which the numerical values of the independent variables have not been fixed. An objective function that defines the value of the design in terms of the independent variables is established.

$$U = U(x_1, x_2, \ldots, x_n) \tag{3.2}$$

Typical objective functions could be cost, weight, reliability, and reducibility. Inevitably, the objective function is subject to certain constraints. Constraints arise from physical laws and limitations or from compatibility conditions on the individual variables. Functional constraints ψ, also called equality constraints, specify relations that must exist between the variables.

$$\begin{aligned}
\psi_1 &= \psi_1(x_1, x_2, \ldots, x_n) \\
\psi_2 &= \psi_2(x_1, x_2, \ldots, x_n) \\
&\vdots \\
\psi_n &= \psi_n(x_1, x_2, \ldots, x_n)
\end{aligned} \tag{3.3}$$

There are no "standard techniques" for optimization in engineering design. How well a technique works depends on the nature of the function represented in the problem. The optimization methods cannot all be covered in a brief chapter, but we shall attempt to discuss a broad range of them as they are applied in engineering design.

As to the optimization by different mathematical methods, we could determine the maximum or minimum values of a function.

Words/Phrases and Expressions

modeling and simulation 建模仿真
thermodynamic 热力学的
static 静力学,静态的
dynamic 动力学,动态的
deterministic 确定性,确定性的
probabilistic 概率性,概率的
iconic 标志性的,图像的
analog 模拟的
symbolic 象征的
wind tunnel 风洞
polymer molecule 聚合物分子

quantitative 定量的
simulation game 仿真游戏
game theory 博弈论
finite element analysis 有限元分析
analytical solution 解析解法
beam 梁
plate 板
algebraic equation 代数方程
mesh 网格
equilibrium 平衡
matrix algebra 矩阵代数
cumbersome 麻烦的
topology 拓扑结构
iterative 迭代的
hole 孔
rib 肋
coequal 相等,同等的
impetus 动力
aluminum 铝
interplay 相互作用
noise 噪声
vibration 振动
optimization 优化
maxima 极大值
minima 极小值
configuration 配置,结构
objective function 目标函数

Chapter 4 Machine Tool

In this chapter, we discuss the technological aspects of the different machining operations as well as the design features of the various machine tools employed to perform those machining operations. In addition, we cover the kinds of shape geometries produced by each operation, the tools used, and the work-holding devices. Moreover, we give special attention to the required workshop calculations, which are aimed at estimating the machining parameters such as cutting speeds and feeds, rate of metal removal, and machining time. But, before doing so, let us talk about machine tools in general.

Machine tools are so designed to drive the cutting tool in order to produce the desired machined surface. For such a goal to be accomplished, a machine tool must include appropriate elements and mechanisms capable of generating the following motions:

(1) A relative motion between the cutting tool cutting and the workpiece in the direction of cutting.

(2) A motion to enable the cutting tool to penetrate into the workpiece until the desired depth of cut is achieved.

(3) A feed motion, which repeats the cutting action every round or every stroke to ensure continuation of the cutting operation.

4.1 The Lathe

Various machining operations can be performed on a conventional engine lathe, for example, cylindrical turning, facing, groove cutting, boring and internal turning, and taper turning. It must be borne in mind, however, that modern computerized numerically controlled lathes have more capabilities and can do other operations, such as contouring.

4.1.1 The Lathe Construction

A lathe is a machine tool used primarily for producing surfaces of revolution and flat edges. Based on their purpose, construction, number of tools that can simultaneously be mounted, and degree of automation, lathes—or, more accurately, lathe-type machine tools—can be classified as follows:

①Engine lathes;
②Toolroom lathes;
③Turret lathes;
④Vertical turning and boring mills;

⑤Automatic lathes;
⑥Special-purpose lathes.

In spite of that diversity of lathe-type machine tools, they all have common features with respect to construction and principle of operation. These features can best be illustrated by considering the commonly used representative type, the engine lathe. Following is a description of each of the main elements of an engine lathe, which is shown in Figure 4.1.

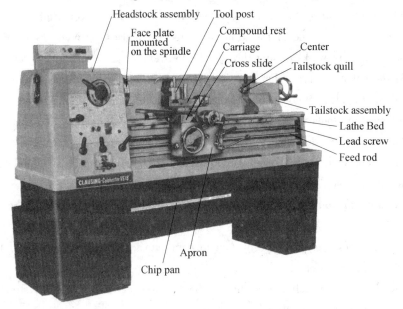

Figure 4.1　The engine lathe

1. Lathe Bed

The lathe bed is the main frame, involving a horizontal beam on two vertical supports. It is usually made of grey or nodular cast iron to damp vibrations and is made by casting. It has guideways to allow the carriage to slide easily lengthwise. The height of the lathe bed should be appropriate to enable the technician to do his or her job easily and comfortably.

2. Headstock

The headstock is fixed at the left-hand side of the lathe bed and includes the spindle, whose axis is parallel to the guideways (the slide surface of the bed). The spindle is driven through the gearbox, which is housed within the headstock. The function of the gearbox is to provide a number of different spindle speeds (usually 6 up to 18 speeds). Some modern lathes have headstocks with infinitely variable spindle speeds, which employ frictional, electrical, or hydraulic drives.

The spindle is always hollow, i.e., it has a through hole extending lengthwise. Bar stocks can be fed through that hole if continuous production is adopted. Also, that hole has a tapered surface to allow mounting a plain lathe center. Such a center is shown in Figure 4.2, it is made of hardened tool steel. The part of the lathe center that fits into the spindle hole has a Morse taper, while the other part of the center is conical, with a 60° apex angle. As is explained later, lathe centers are used for mounting long workpieces. The outer surface of the spindle is threaded to allow mounting of a chuck, a face plate, or the like.

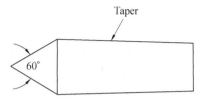

Figure 4.2   A sketch of a plain lathe center

3. Tailstock

The tailstock assembly consists basically of three parts, its lower base, an intermediate part, and the quill. The lower base is a casting that can slide on the lathe bed along the guideways, and it has a clamping device to enable locking the entire tailstock at any desired location, depending upon the length of the workpiece. The intermediate part is a casting that can be moved transversely to enable alignment of the axis of the tailstock with that of the headstock. The third part, the quill, is a hardened steel tube, which can be moved longitudinally in and out of the intermediate part as required. This is achieved through the use of a handwheel and a screw, around which a nut fixed to the quill is engaged. The hole in the open side of the quill is tapered to enable mounting of lathe centers or other tools like twist drills or boring bars. The quill can be locked at any point along its travel path by means of a clamping device.

4. The Carriage

The main function of the carriage is mounting of the cutting tools and generating longitudinal and/or cross feeds. It is actually an H-shaped block that slides on the lathe bed between the headstock and tailstock while being guided by the V-shaped guideways of the bed. The carriage can be moved either manually or mechanically by means of the apron and either the feed rod or the lead screw.

The apron is attached to the saddle of the carriage and serves to convert the rotary motion of the feed rod (or lead screw) into linear longitudinal motion of the carriage and, accordingly, the cutting tool—i.e., it generates the axial feed. The apron also provides powered motion for the cross slide located on the carriage. Usually, the tool post is mounted on the compound rest, which is in turn mounted on the cross slide.

4.1.2   The Lathe Tool

The shape and geometry of the lathe tools depend upon the purpose for which they are employed. Turning tools can be classified into two main groups, namely, external cutting tools and internal cutting tools. Each of these two groups includes the following types of tools: turning tools, facing tools, cutoff tools, thread-cutting tools, and form tools. The types of the internal cutting tools are similar to those of the external cutting tools. They include tools for rough turning, finish turning, thread cutting, and recess machining.

4.2 Drilling Machine

Drilling involves producing through or blind holes in a workpiece by forcing a tool, which rotates around its axis, against the workpiece. Consequently, the range of cutting from that axis of rotation is equal to the radius of the required hole.

4.2.1 Construction and Accessories

The standard drill press (Figure 4.3) has a base which may be used as a table for mounting work when the maximum distance between the spindle and the table is desired. This base table is different from the regular work table in that it acts as the support for the column and is in a fixed position relative to the spindle. The regular work table is clamped to the column so that it can be positioned vertically along the column or radially about the column. Some of the more complex drill presses have provisions for moving table laterally and longitudinally as well as vertically along the column.

The drill head and power unit are mounted at the top of the column. The classification of bench versus floor drill press is essentially related to the length of the column.

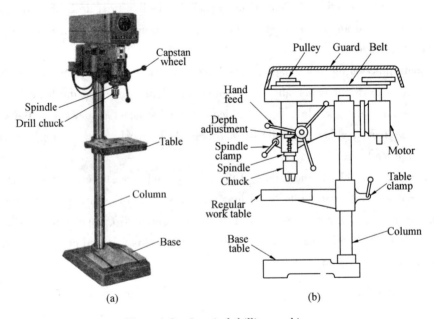

Figure 4.3 A typical drilling machine

4.2.2 Drill Bits

The drill bits may be used, or adapted for use, on the rotary type of power tools. They are manufactured with straight, tapered, or special kinds of shanks for quick locking. The shank is usually soft to facilitate holding. The cutting portion of the drill is made from hardened high-carbon high-speed or cobalt-alloy steels. The drill bit may also be made from carbide or fitted with carbide inserts. The flutes may be straight or have long leads or short leads. The helixes of most manufactured drills are right-hand. When viewed from the cutting edge, the helix winds around the drill in a clockwise direction. Thus the drill must rotate counterclockwise when viewed from the cutting end.

1. Twist Drill

The twist drill is the most common type of drill. It has two cutting edges and two helical flutes that continue over the length of the drill body, as shown in Figure 4.4. The drill also consists of a neck and a shank that can be either straight or tapered. In the latter case, the shank is fitted by the wedge action into the tapered socket of the spindle and has a tang, which goes into a slot in the spindle socket, thus acting as a solid means for transmitting rotation. On the other hand, straight-shank drills are held in a drill chuck that is, in turn, fitted into the spindle socket in the same way as tapered shank drills. Twist drills are usually made of high-speed steel, although carbide-tipped drills are also available.

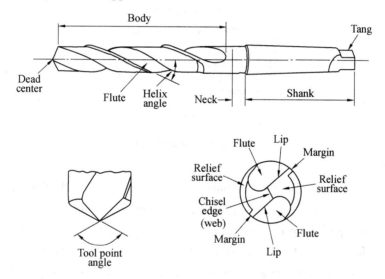

Figure 4.4 The twist drill

2. Core Drills

A core drill consists of the chamfer, body, neck, and shank, as shown in Figure 4.5. This type of drill may have either three or four flutes and an equal number of margins, which ensure superior guidance, thus resulting in high machining accuracy. It can also be seen in Figure 4.6 that a core drill has flat end. The chamfer can have three or four cutting edges, or lips, and the lip angle may vary between 90° and 120°. Core drills are employed for enlarging previously made holes and not for originating holes. This type of drill is characterized by greater productivity, high machining accuracy, and superior quality of the drilled surfaces.

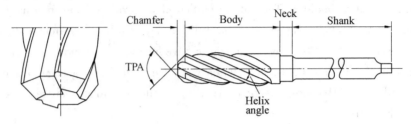

Figure 4.5　The core drill

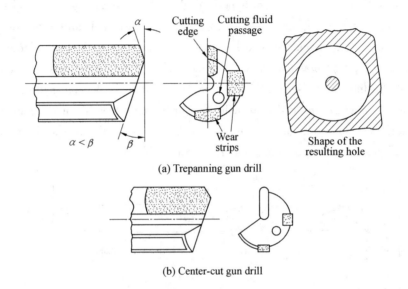

(a) Trepanning gun drill

(b) Center-cut gun drill

Figure 4.6　The gun drill

3. Gun Drills

Gun drills are used for drilling deep holes. All gun drills are straight-fluted, and each has a single cutting edge. A hole in the body acts as a conduit to transmit coolant under considerable pressure to the tip of the drill. As can be seen in Figure 4.6, there are two kinds of gun drills, namely, the center-cut gun drill used for drilling blind holes and the trepanning drill. The latter has a cylindrical groove at its center, thus generating a solid core, which guides the tool as it proceeds during the drilling operation.

4. Spade Drills

Spade drills are used for drilling large holes of 3.5 in (90 mm) or more. Their design results in a marked saving in cost of the tool as well as a tangible reduction in its weight, which facilitates its handling. Moreover, this type of drill is easy to grind. Figure 4.7 indicates a spade drill.

5. Saw-type Cutters

Saw-type cutters, illustrated in Figure 4.8, are used for cutting large holes in thin metal.

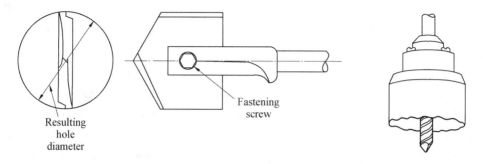

Figure 4.7　The spade drill　　　　　　　　Figure 4.8　The saw-type drill

6. Drills Made in Combination with Other Tools

An example of a combination drill is a tool that involves a drill and a tap together. Step drills and drill and countersink tools are also sometimes used in industrial practice.

4.2.3　Cutting Speeds and Feeds in Drilling

It is clear that the cutting speed varies along the cutting edge. It is always maximum at the periphery of the tool and is equal to zero on the tool axis. Nevertheless, we consider the maximum speed, since it is the one that affects the tool wear and the quality of the machined surface. The maximum speed must not exceed the permissible cutting speed, which depends upon the material of the workpiece as well as the material of the cutting tool. Data about permissible cutting speeds in drilling operations can be found in handbooks. The rotational speed of the spindle can be determined from the following equation:

$$N = \frac{C.S.}{\pi D} \qquad (4.1)$$

where　N——the rotational speed of the spindle(r/min);

　　　　D——the diameter of the drill(ft);

　　　　C.S.——the permissible cutting speed.

In drilling operations, feeds are expressed as inches or millimeters per revolution. Again, the appropriate value of feed to be used depends upon the metal of the workpiece and the drill material and can be found in handbooks. Whenever the production rate must be increased, it is advisable to use higher feeds rather than to increase the cutting speed.

4.3　Milling

4.3.1　Introduction

Milling is a machining process that is carried out by means of a multiedge rotating tool known as a milling cutter. In this process, metal removal is achieved through combining the rotary motion of the milling cutter and linear motions of the workpiece simultaneously. Milling operations are employed in producing flat, contoured and helical surfaces as well as for thread- and gear-cutting operations.

Each of the cutting edges of a milling cutter acts as an individual single-point cutter when it engages with the workpiece metal. Therefore, each of those cutting edges has appropriate rake and relief angles. Since only a few of the cutting edges are engaged with the workpiece at a time, heavy cuts can be taken without adversely affecting the tool life. In fact, the permissible cutting speeds and feeds for milling are three to four times higher than those for turning or drilling. Moreover, the quality of the surfaces machined by milling is generally superior to the quality of surfaces machined by turning, shaping, or drilling.

A wide variety of milling cutters is available in industry. This, together with the fact that a milling machine is a very versatile machine tool, makes the milling machine the backbone of a machining workshop.

As far as the direction of cutter rotation and workpiece feed are concerned, milling is performed by either of the following two methods.

1. Up Milling(Conventional Milling)

In up milling the workpiece is fed against the direction of cutter rotation, as shown in Figure 4.9(a). As we can see in that figure, the depth of cut (and consequently the load) gradually increases on the successively engaged cutting edges. Therefore, the machining process involves no impact loading, thus ensuring smoother operation of the machine tool and longer tool life. The quality of the machined surface obtained by up milling is not very high. Nevertheless, up milling is commonly used in industry, especially for rough cuts.

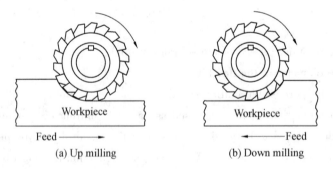

Figure 4.9　Milling methods

2. Down Milling (Climb Milling)

As can be seen in Figure 4.9(b), in down milling the cutter rotation coincides with the direction of feed at the contact point between the tool and the workpiece. It can also be seen that the maximum depth of cut is achieved directly as the cutter engages with the workpiece. This results in a kind of impact, or sudden loading. Therefore, this method cannot be used unless the milling machine is equipped with a backlash eliminator on the feed screw. The advantages of this method include higher quality of the machined surface and easier clamping of workpieces, since the cutting forces act downward.

4.3.2 Types of Milling Machines

There are several types of milling machines employed in industry. They are generally classified based on their construction and design features. They vary from the common general-purpose types to duplicators and machining centers that involve a tool magazine and are capable of carrying out many machining operations with a single workpiece setup. Following is a survey of the milling machine types commonly used in industry.

1. Plain Horizontal Milling Machine

The construction of the plain horizontal milling machine is very similar to that of the universal milling machine shown in Figure 4.10, except that the machine table cannot be swiveled. Plain milling machines usually have column and knee-type construction and also have three table motions, i.e., longitudinal, transverse, and vertical. The milling cutter is mounted on a short arbor, which is, in turn, rigidly supported by the overarm of the milling machine.

2. Vertical Milling Machine

As the name vertical milling machine suggests, the axis of the spindle that holds the milling cutter is vertical. Table movements are generally similar to those of plain horizontal milling machines. However, an additional rotary motion is sometimes provided for the table to enable machining helical and circular grooves. The cutters used with vertical milling machines are almost always of the end mill type. Figure 4.11 illustrates the construction and design features of a vertical milling machine.

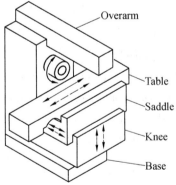
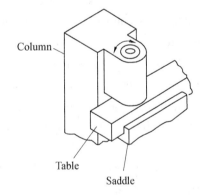

Figure 4.10　The horizontal milling machine　　Figure 4.11　The vertical milling machine

3. Bed-Type Milling Machine

High-production milling operations are carried out on bed-type milling machines (Figure 4.12) which have less versatility but more rigidity. The general characteristics of this type of machine are that the height of the table is fixed and the adjustments for height are made with the spindles. This height adjustment is made by moving the spindle head up or down. On some machines, the transverse adjustment may be made with the table. On others, the spindles may be adjusted in or out. In such cases, machine utilization is increased by placing a work fixture at either end so that one fixture is loaded while the work in the other is being machined.

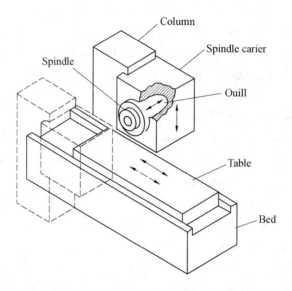

Figure 4.12　The bed-type milling machine

4. Universal Milling Machine

The construction of a universal milling machine is similar to that of the plain milling machine, except that it is more accurate and has sturdier frame, and its table can be swiveled with an angle up to 50°. Universal milling machines are usually equipped with an index or dividing head, which allows cutting of gears and cams.

5. Duplicators

A duplicator is sometimes referred to as a copy milling machine because it is capable of reproducing an exact replica of a model. The machine has a stylus that scans the model, at which time counterpoints on the part are successively machined. Duplicators were used for the production of large forming dies for the automotive industry, where models made of wood, plaster of Paris, or wax were employed. Duplicators are not commonly used in industry nowadays because they have been superseded by the CAD/CAM systems.

6. Machining Centers

Machining centers are multipurpose NC machines that are capable of performing a number of different machining processes at a time. A machining center has a tool magazine in which many tools are held. Tool changes are automatically carried out, and so are functions such as coolant on or off. Machining centers are, therefore, highly versatile and are employed to perform a number of machining operations on a workpiece with a single setup. Parts having intricate shapes can easily be produced with high accuracy and excellent repeatability.

4.3.3 Types of Milling Cutters

There is a wide variety of milling cutter shapes. Each of them is designed to perform effectively a specific milling operation. Generally, a milling cutter can be described as a multiedge cutting tool having the shape of a solid of revolution, with the cutting teeth arranged either on the periphery or on an end face or on both. Following is a quick survey of the commonly used types of milling cutters.

1. Plain Milling Cutter

A plain milling cutter is a disk-shaped cutting tool that may have either straight or helical teeth, as shown in Figure 4.13(a). This type is always mounted on horizontal milling machines and is used for machining flat surfaces.

2. Face Milling Cutter

A face milling cutter is also used for machining flat surfaces. It is bolted at the end of a short arbor, which is in turn mounted on a vertical milling machine. Figure 4.13(b) indicates a milling cutter of this type.

3. Plain Metal Slitting Saw

Figure 4.13(c) indicates a plain metal slitting saw cutter. We can see that it actually involves a very thin plain milling cutter.

4. Side Milling Cutter

A side milling cutter is used for cutting slots, grooves, and splines. As we can see in Figure 4.13(d), it is quite similar to the plain milling cutter, the difference being that this type has teeth on the sides. As was the case with the plain cutter, the cutting teeth can be straight or helical.

5. Angle Milling Cutter

An angle milling cutter is employed in cutting dovetail grooves, ratchet wheels, and the like. Figure 4.13(e) indicates a milling cutter of this type.

6. T-slot Cutter

As shown in Figure 4.13(f), a T-slot cutter involves a plain milling cutter with an integral shaft normal to it. As the name suggests, this type is used for milling T-slots.

7. End Mill Cutter

End mill cutters find common applications in cutting slots, grooves, flutes, splines, pocketing work, and the like. Figure 4.13(g) indicates an end mill cutter. The latter is always mounted on a vertical milling machine and can have two or four flutes, which may be either straight or helical.

8. Form Milling Cutter

The teeth of a form milling cutter have a certain shape, which is identical to the section of the metal to be removed during the milling operation.

Examples of this type include gear cutters, gear hobs, convex and concave cutters, and

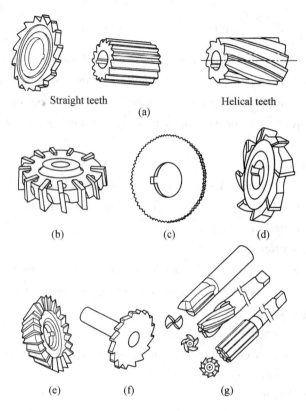

Figure 4.13 Types of milling cutters

the like. Form milling cutters are mounted on horizontal milling machines, as is explained later when we discuss gear cutting.

4.4 Shaping and Planing Operations

Planing, shaping, and slotting are processes for machining horizontal, vertical, and inclined flat surfaces, slots, or grooves by means of a lathe-type cutting tool. In all these processes, the cutting action takes place along a straight line. In planing, the workpiece (and the machine bed) is reciprocated, and the tool is fed across the workpiece to reproduce another straight line, thus generating a flat surface. In shaping and slotting, the cutting tool is reciprocated and the workpiece is fed normal to the direction of tool motion. The difference between the latter two processes is that in shaping, the tool path is horizontal, it is vertical in slotting. Shapers and slotters can be employed in cutting external and internal keyways, gear racks, dovetails, T-slots, and the like. In fact, shapers and planers have virtually become obsolete, as most of planing and shaping operations have been replaced by more productive processes such as milling, broaching, or abrasive machining. The use of shapers and planers is now limited to the machining of large beds of machine tools and the like.

In all these three processes, there are successive alternating cutting and idle return strokes. The cutting speed is, therefore, the speed of the tool (or the workpiece) in the direction of cutting during the working stroke. The cutting speed may be either constant throughout the working stroke or variable, depending upon the design of the shaper or planer. We now discuss the construction and operation of the most common types of shapers and planers.

4.4.1 Shaper

1. Horizontal Push-Cut Shapers

(1) Construction.

As can be seen in Figure 4.14, this type of machine tool consists of a frame that houses the speed gearbox and the quick-return mechanism, which transmits power from the motor to the ram and the table. The ram travel is the primary motion that produces a straight-line cut in the working stroke, whereas the intermittent cross travel of the table is responsible for the cross feed. The tool head is mounted at the front end of the ram and carries the clapper box tool holder. The tool holder is pivoted at its upper end to allow the tool to rise during the idle return stroke in order not to ruin the newly machined surface. The tool head can be swiveled to enable machining inclined surfaces.

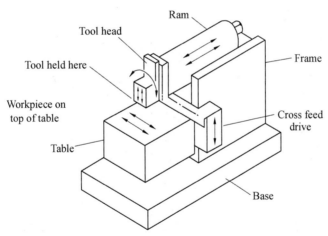

Figure 4.14 Horizontal push-cut shaper

The workpiece can either be bolted directly to the machine table or held in a vise or any other suitable fixture. The cross feed of the table is generated by a ratchet and pawl mechanism, which is driven through the quick-return mechanism (i.e., the crank and the slotted arm). The machine table can be raised or lowered by means of a power screw and a crank handle. It can also be swiveled in a universal shaper.

(2) Quick-return mechanism.

The quick-return mechanism (Figure 4.15) involves a rotating crank that is driven at a uniform angular speed and an oscillating slotted arm that is connected to the crank by a sliding block. The working stroke takes up an angle (of the crank revolution) that is larger than that of the return stroke. Since the angular speed of the crank is constant, it is obvious that the time taken by the idle return stroke is less than that taken by the cutting stroke. In fact, it is the main function of the quick-return mechanism to reduce the idle time during the machining operation to a minimum.

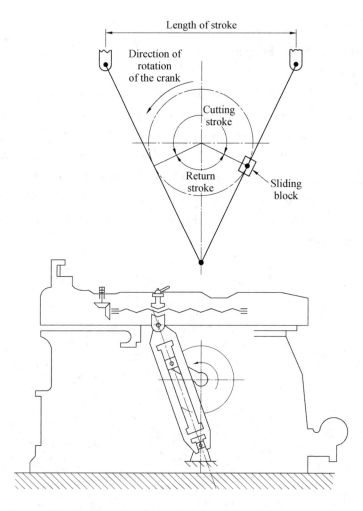

Figure 4.15　Details and working principles of the quick-return mechanism

2. Vertical Shaper

The vertical shaper is similar in construction and operation to the push-cut shaper, the difference being that the ram and the tool head travel vertically instead of horizontally. Also, in this type of shapers, the workpiece is mounted on a round table, which can have a rotary feed whenever desired to allow machining of curved surfaces, e. g. , spiral grooves. Vertical shapers, which are sometimes referred to as slotters, are used in internal cutting. Another type of vertical shapers is known as a keyseater because it is specially designed for cutting keyways in gears, cams, pulleys, and the like.

4.4.2 Planer

A planer is a machine tool that does the same work as the horizontal shaper but on workpieces that are much larger than those machined on shapers (Figure 4.16). Although the designs of planers vary, most common are the double-housing and open-side construction. In a double-housing planer, two vertical housings are mounted at the sides of the long heavy bed. A cross rail, which is supported at the top of these housings, carries the cutting tools. The machine table (while in operation) reciprocates along the guideways of the bed and has T-slots in its upper surface for clamping the workpiece. In this type of planer, the table is powered by a variable-speed DC motor through a gear drive. The cross rail can be raised or lowered as required, and the inclination of the tools can be adjusted as well. In an open-side planer, there is only one upright housing at one side of the bed. This latter construction provides more flexibility when wider workpieces are to be machined.

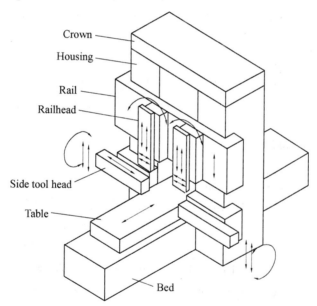

Figure 4.16 Horizontal push-cut shaper

4.4.3 Planing and Shaping Tools

Planing and shaping processes employ single-point tools of the lathe type, though heavier in construction. They are made of either high-speed steel or carbon tool steel with carbide tips. In the latter case, the machine tool should be equipped with an automatic lifting device to keep the tool from rubbing the workpiece during the return stroke, thus eliminating the possibility of breakage or chipping of the carbide tips.

The cutting angles for those tools depend upon the purpose for which the tool is to be used and the material being cut. The end relief angle does not usually exceed 4°, whereas the side relief varies between 6° and 14°. The side rake angle also varies between 5° (for cast iron) and 15° (for medium carbon steel).

4.5 Grinding

Grinding is a manufacturing process that involves the removal of metal by employing a rotating abrasive wheel. The latter simulates a milling cutter with an extremely large number of miniature cutting edges. Generally, grinding is considered to be a finishing process that is usually used for obtaining high-dimensional accuracy and better surface finish. Grinding can be performed on flat, cylindrical, or even internal surfaces by employing specialized machine tools, which are referred to as grinding machines. Obviously, grinding machines differ in construction as well as capabilities, and the type to be employed is determined mainly by the geometrical shape and nature of the surface to be ground, e.g., cylindrical surfaces are ground on cylindrical grinding machines.

4.5.1 Grinding Types

There are five major types of grinding operations: cylindrical (center or centerless), internal or hole, surface, tool and cutter, and abrasive belt grinding for precision and surface-finishing operations.

1. Cylindrical Grinding

In cylindrical grinding, the workpiece is held between centers during the grinding operation, and the wheel rotation is the source and cause for the rotary cutting motion, as shown in Figure 4.17. In fact, cylindrical grinding can be carried out by employing any of the following methods.

(1) The transverse method, in which both the grinding wheel and the workpiece rotate and longitudinal linear feed is applied to enable grinding of the whole length. The depth of cut is adjusted by the cross feed of the grinding wheel into the workpiece.

(2) The plunge-cut method, in which grinding is achieved through the cross feed of the grinding wheel and no axial feed is applied. As you can see, this method can be applied only when the surface to be ground is shorter than the width of the grinding wheel used.

(3) The full-depth method, which is similar to the transverse method except that the grinding allowance is removed in a single pass. This method is usually recommended when grinding short rigid shafts.

2. Centerless Grinding

Centerline grinding is a rapid, economical production operation. When grinding on centers, the work revolves on fixed centers; whereas in centerless grinding the diameter is determined by the periphery of the workpiece. In centerless grinding, the work is supported by three machine elements: the grinding wheel, the regulating wheel, and the work-rest blade, which is

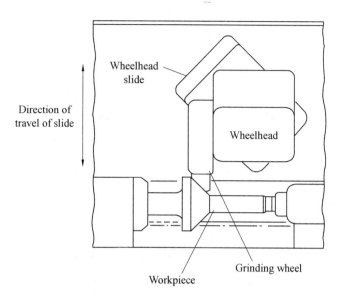

Figure 4.17 Block digram of a typical cylindrical grinder

beveled to push the workpiece toward the regulating wheel (Figure 4.18). The rubber-bonded regulating wheel controls the rotational speed of the workpiece, and its rate of feed is determined by the angle of the horizontal axis of the regulating wheel with respect to that of the grinding wheel. Note that the faces of both wheels are always parallel.

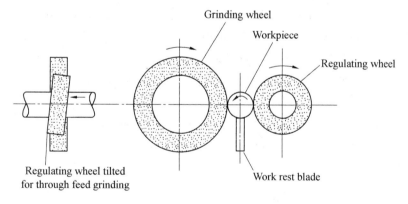

Figure 4.18 Schematic sketch showing the principle of centerless grinding

There are two types of centerless grinding: through feed and infeed (Figure 4.19). In through feed grinding the workpiece passes completely through the grinding zone and exits on the opposite side. Infeed grinding is similar to a plunge cut when grinding on centers. In this case the length of the section to be ground is limited by the width of the grinding wheel.

3. Internal Grinding

There are three types of internal grinding machines:

(1) Machines in which the workpiece rotates slowly and the wheel spindle rotates and reciprocates the length of the hole.

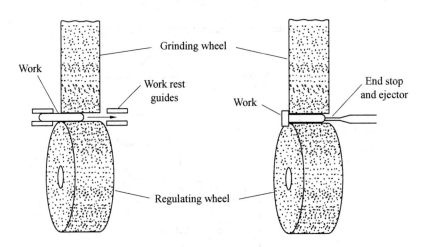

Figure 4.19 Through feed and infeed centerless grinding

(2) Machines in which the work rotates and reciprocates while the wheel spindle rotates only.

(3) Machines in which the wheel spindle rotates and has a planetary motion while the workpiece reciprocates.

4.5.2 Grinding Wheels

Grinding wheels are composed of abrasive grains having similar size and a binder. The actual grinding process is performed by the abrasive grains. Pores between the grains within the binder enable the grains to act as separate single-point cutting tools. These pores also provide space for the generated chips, thus preventing the wheel from clogging. In addition, pores assist the easy flow of coolants to enable efficient and prompt removal of the heat generated during the grinding process.

Grinding wheels are identified based on their shape and size, kind of abrasive, grain size, binder, grade (hardness), and structure.

4.6 Sawing Operations

Parting or cutoff operations can be performed on different machine tools such as engine lathes, milling machines, and grinding machines. Nevertheless, when cutting off is a basic operation in a large-volume production line, special sawing machines are required to cope with that production volume.

4.6.1 Sawing Operations

The cutting tool may take different forms, depending upon the type of sawing machine used. The tool can be a blade, a circular disk, or a continuous band. However, all these tools are multiedge with several cutting edges (i. e., teeth) per inch. As can be seen in Figure 4.20, teeth can be straight, claw, or buttress. Each tooth, irrespective of its form, must have a rake and a relief angle. Also, teeth are offset in order to make the kerf wider than the thickness of each individual tooth. This facilitates easy movement of the saw blade in the kerf, thus reducing the friction and the generated heat. The maximum thickness is usually referred to as the saw set and is apparently equal to the width of the resulting kerf. When selecting the cutting speed and the number of teeth per inch, several factors have to be taken into consideration, such as the cutting tool material, the material of the workpiece, the tooth form, and the type of lubricant (coolant) used.

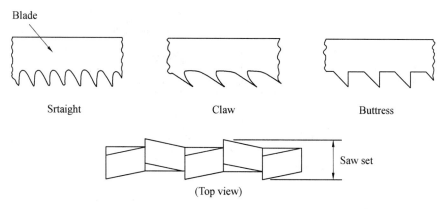

Figure 4.20 Types of sawing teeth

4.6.2 Sawing Machines

Sawing machines differ in shape, size, and construction, depending upon the purpose for which they are to be used. Nevertheless, they can be classified into three main groups.

1. Reciprocating Saw

In the reciprocating saw, a relatively large hacksaw blade is mechanically reciprocated. Depending upon the construction of the saw, the cutting blade may be either horizontal or vertical. We can see that each cycle has a working (cutting) stroke as well as an idle stroke. Consequently, this type is considered to be a low-productivity saw and is used only in small shops with low to moderate production volume.

2. Circular Saw

The cutting tool in a circular saw is a circular disk, with the cutting teeth uniformly arranged on its periphery. It looks like the slitting cutter used with milling machines. Although it is highly efficient, it can only be used for parting or cutoff operations of bar stocks or rolled sections.

3. Band Saw

The highly flexible and versatile band saw employs a continuous-band sawing blade. As can be seen in Figure 4.21, the band saw blade is mounted on two pulleys, one of which is the source of power and rotation. Each machine has a flash welding attachment in order to weld the edges of the band saw blade together after adjusting the length, thus forming a closed band. Band saws can be used in contouring as well as for large-volume cutoff operations. Loading and unloading of the bar stock as well as length adjustment are done automatically (by special attachments in the latter case).

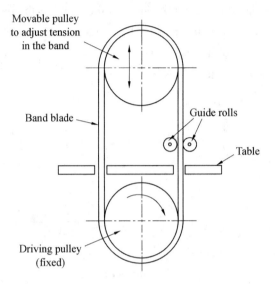

Figure 4.21　Basic idea of a band saw

4.7　Broaching Operations

Broaching provides high repetitive accuracy (applicable to production of large numbers of parts of close tolerances and fine finish) and close dimensional relationship of several surfaces broached simultaneously. Broaching is 15 to 25 times faster than other competitive machining methods. The process can be used to accurately produce internal and external surfaces that are difficult to machine by other methods.

The principal disadvantage of broaching is the high cost of special broaching tools. This cost usually does not permit the process to be used when production requirements are low. It cannot be employed economically for the removal of large amounts of stock (more than 1/4 in(1 in=25.4 mm)). Lastly, the process has application only on unobstructed surfaces that permit the pass of the broach through the workpiece.

Broaching's principal applications are for the production of almost any desired external or internal contour. This includes flat, round, and irregular external surfaces, round and square holes, splines, keyways, rifling, and gear teeth.

Broaching is a metal-removing operation in which a multiedge cutting tool, like that shown in Figure 4.22, is used. In this operation, only a thin layer or limited amount of metal is removed. Broaching is commonly used to generate internal surfaces or slots, which are very difficult to produce otherwise. However, it can also be used for producing intricate external surfaces that require tight tolerances.

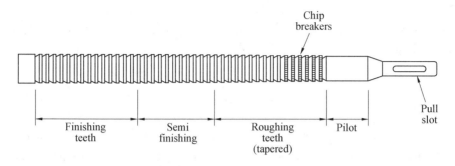

Figure 4.22　A broaching tool

4.7.1　Broaching Machines

A broaching machine simply includes a sturdy frame (or bed), a device for locating and clamping the workpiece, the cutting tool, and a means for moving the cutting tool (or the workpiece). Following is a quick survey of the commonly used types of broaching machines:

(1) Pull-type machines, in which the broaching tool is withdrawn through the initial hole in the workpiece, which is tightly clamped;

(2) Push-type machines, in which the broaching tool is pushed to generate the required surface;

(3) Surface broaching machines, where either the tool or the workpiece move to generate the desired surface;

(4) Continuous broaching machines, where the workpiece moves continually over a fixed broaching tool, the path can be either straight or circular.

4.7.2　Advantages and Limitations of Broaching Operations

It is of importance to know the advantages and limitations of broaching in order to make full use of the potentials of that operation. The advantages include the high cutting speed and high cycling time, the close tolerances and superior surface quality that can easily be achieved, and the fact that both roughing and finishing are combined in the same stroke of the broaching tool. Nevertheless, this operation can be performed only on through holes or external surfaces and cannot be carried out on blind holes. In addition, broaching involves only light cuts and, therefore, renders itself unsuitable for operations where the amount of metal to be removed is relatively large. Finally, the high cost of broaching tools and machines, together with the expensive fixturing, makes this operation economically unjustifiable unless a large number of products are required.

4.8 Precision and Surface-Finishing Operations

After secondary machining, either the precision or the surface finish may require further refinement. In this case, fine abrasive machining is required. If both precision and finish are required, lapping or honing are indicated. If surface finish alone is desired, then fine abrasive blasting, buffing, tumbling, vibratory finishing, or polishing may be sufficient.

4.8.1 Lapping

Lapping is an abrading process that results in a wearing down of the ridges and high spots on a machined surface, leaving the valleys and a random array of fine scratches. Lapping is used to obtain fine dimensional accuracy, to correct minor imperfections in shape, to secure a fine surface finish, and to obtain a fit between two mating surfaces. In the lapping process a fine abrasive, carried in a light oil medium, is supplied to the work surface by a reciprocating type of motion that occurs in an ever-changing path. The resultant abrading action between the work surface and the softer lap such as brass, lead, cast iron, wood, or leather results in the removal of the high spots on the work surface. The cutting action occurs because the soft surface of the lap becomes charged with a layer of abrasive particles that provide a myriad of cutting edges while the lap surface maintains the proper contour.

The service life of lapped surfaces is usually substantially increased over that of the same parts when mated without lapping. Mating gears and worms are lapped to remove imperfections resulting from heat treatment or prior machining. Piston rings and gage blocks are customarily made parallel and to high precision by lapping.

The lapping process can be carried out economically by using the principles of mechanization and automation. If an eccentric spider is used to carry the parts that are to be made parallel in an ever-changing path between two rotating, parallel plates with a recirculating abrasive slurry, rapid and precise lapping will occur. Grooves in a checkerboard pattern in the lapping plate facilitate the delivery of the slurry and help flush away the surface particles. Wet lapping has been shown to be at least six times faster than dry lapping and there is no heat buildup.

Manual lapping is used to bring gage blocks to their final stage of dimensional accuracy and parallelism—a finish of 1 to 2 μin and a tolerance as small as ±0.000 001 in.

4.8.2 Honing

Honing may be defined as a production method for finishing internal or external surfaces revolution after machining or grinding. Honing is usually accomplished by reciprocating several spring-loaded sticks of a fine abrasive material over the surface of the rotating workpiece. Honing provides a fine, finished texture in holes and on the exterior surface as well. Bored holes are frequently reamed or honed to obtain final dimensional accuracy and a characteristic figure-of-eight pattern that retains oil on the surface over an extended period of time. This is particularly fortunate in the case of internal-combustion engines, especially during the breaking-in period. Typical honed cylinders are internal-combustion engine cylinders, bearings, gun barrels, ring gages, piston pins, and shafts. Honing is a cutting operation and rarely exceeds a removal of more than 0.001 in of stock on a side. Boring and beaming should precede honing to assure surface shape and location. Although a 1-μm surface finish is possible, an 8-to 10-μin finish is more economical and probably just as good.

Honing machines can be either horizontal or vertical. The former are used for long holes such as cannon or rifle bores or in small models for ring gages. The latter type are more common, ranging from those capable of handling all eight bores of a V-8 engine simultaneously to large machines with an 8-ft (1 ft = 304.8 mm) stroke capable of handling bores up to 30 in in diameter.

A typical internal hone consists of a framework for supporting the abrasive stones, which are mounted and internally spring-loaded so that when in the hole they can expand to make contact with the surface of the bore. A cutting fluid is always used to flush away the abraded particles, to improve the finish of the workpiece, and to keep the stones free cutting.

4.8.3 Superfinishing

Superfinishing is a proprietary name, but the process produces the ultimate in refinement of surfaces. Although it is similar to honing, its principles are basically different. A large area of abrasive is used, so that uneven projections and wavy surfaces are removed through a clean cut. Since the abrasive is moved in many directions, it is self-adjusting and becomes automatically a master shape that will correct the work surface. In this respect it is similar to lapping, but the surface is not charged with lapping material. The process does not remove major amounts of material (maximum of 0.0001 in) in the average production job. The abrasive may be stones, wheels, or belts supported by master platens. Superfinishing is in common use on automobile parts, and automatic machines have been developed to automate this process, which is effective in removing chatter marks and amorphous material until a true surface of the base metal remains. The abrasive is loosely bonded so that each reciprocating stone quickly wears to the contour of the part, but it is large enough so that a representative surface is encountered. Thus, the stone bridges and equalizes a number of defects simultaneously and corrects the surface to an average profile. As the surface becomes smoother the unit pressure decreases until the stone rides on a fluid film and cuts very little. A 1-in wide ground surface may be refined to a 3-μin finish in less than 1 min.

Words/Phrases and Expressions

machine tool 机床
lathe 车床

cylindrical turning 圆柱车削
facing 端面
groove cutting 切槽
boring and internal turning 镗孔和内孔车削
taper turning 锥度车削
engine lathe 普通车床
toolroom lathe 工具车床
turret lathe 转塔车床
vertical turning and boring mill 立式镗铣床
automatic lathe 自动车床
special-purpose lathe 专用车床
bed 床身
guideway 导轨
headstock 主轴箱
gearbox 变速箱
tailstock 尾座
carriage 溜板箱
drilling 钻削
drill bit 钻头
twist drill 麻花钻
core drill 空心钻
gun drill 枪钻
spade drill 扁钻
saw-type cutter 锯齿钻
milling 铣削
multiedge 多边
up milling(conventional milling)逆铣
down milling (climb milling)顺铣
plain horizontal milling machine 平面卧式
vertical milling machine 立式
bed-type milling machine 台式铣床
universal milling machine 万能铣床
duplicator 复制铣床
machining center 加工中心
shaper 牛头刨床
planer 龙门刨床
slotter 开槽机
grinding 磨削

cylindrical grinding 外圆磨削
centerless grinding 无心磨削
internal grinding 内圆磨削
sawing 锯切
kerf 切口
broaching 拉削
precision 精密
surface-finishing 表面处理
lapping 研磨
honing 珩磨
superfinishing 超精加工

Chapter 5 CNC System

5.1 Introduction

Numerical Control (NC) is the technique of giving instructions to a machine in the form of a code which consists of numbers, letters of the alphabet, punctuation marks and certain other symbols. The machine responds to this coded information in a precise and ordered manner to carry out various machining functions. These functions may range from the positioning of the machine spindle relative to the workpiece (the most important function), to controlling the speed and direction of spindle rotation, tool selection, on/off control of coolant flow, and so on.

Instructions are supplied to the machine as blocks of information. A block of information is a group of commands sufficient to enable the machine to carry out one individual machining operation. A set of instructions forms an NC program. When the instructions are organised in a logical manner they direct the machine tool to carry out a specific task—usually the complete machining of a workpiece or "part". It is thus termed a part program. Such a part program may be utilized, at a later date, to produce identical results over and over again.

5.2 NC Machine Tools

Historically, machine tools were controlled with handles, or levers, by an individual. An electric motor supplied the power to move a workpiece either linearly or in rotary fashion. The power was also directed to a mechanism that rotated and/or moved the workpiece. In the late 19th century and the early 20th century, automation was in the form of production milling machines that used cams and preset stops to control the motion of the table. Production lathes were characterized by capstans. Other forms of production machines were turret lathes, copy machines, tracers, and so on. The controls were mechanical, electrical, hydraulic, or a combination of these. They provided the basis for mass production, interchangeability, repeated dimensional accuracy, higher production rates, reduced labor and labor costs, and so on.

Automatic control of NC machine tools relies on the presence of the part program in a form that is external to the machine itself. The NC machine does not possess any "memory" of its own and as such is only capable of executing a single block of information, fed to it, at a time. For this reason, part programs are normally produced, and stored, on punched tape. The numerically controlled (NC) machine tool may be open-loop, closed-loop, or a combination of open- and closed-loop.

5.3 Computer Numerical Control

Computer Numerical Control (CNC) retains the fundamental concepts of NC but utilizes a dedicated stored-program computer within the machine control unit (Figure 5.1). CNC is largely the result of technological progress in microelectronics (the miniaturisation of electronic components and circuitry), rather than any radical departure in the concept of NC.

Figure 5.1 Super precision CNC machines such as this Super-Mimi Mill, have resolutions below 0.0001 in

In computer numerical control (CNC), the memory is a computer. Essentially, a computer is the control system applied to CNC machinery. This makes it possible to store information in the computer instead of punching holes in a tape. In this manner much more information can be stored and more easily retrieved. Should conditions change, it is also easier to access the computer memory and make changes than it is to make corrections and cut a new tape.

CNC control units, like the computers on which they are based, operate according to a stored program held in computer memory. This means that part programs are now able to become totally resident within the memory of the control unit, prior to their execution. No longer do the machines have to operate on the "read-block/execute-block" principle. This eliminates the dependency on slow, and often unreliable, tapes and tape reading devices—probably the weakest link in the chain. Programs can, of course, still be loaded into the CNC machine via punched tape, but only one pass is necessary to read the complete part program into the memory of the control unit.

5.4 CNC Machine Technical Terminologies

1. Axis Drive Mechanics

We'll investigate the mechanics and how the kinds of axis make them move. We'll learn how they can make a near perfect surface when cutting into and out of changing material loads. Even with varying excess from one part to the next, every part comes off the machine a near perfect duplicate of the last.

2. Machine Variations and Evolution

While today's machines are far superior to anything in the past, it's a sure bet they aren't the ultimate. So, our goal is an understanding of the drive behind machine evolution such that the future is easier to see. Reading the time line carefully, you might foresee a big change.

3. Programs

We'll also examine program creation and data management. This chapter isn't about writing programs. It's about where they come from and how they are managed.

4. A CNC Axis Drive

The biggest difference between a manual machine and a modern industrial CNC is the way it moves its axes. A CNC axis shares a lot of similarities with using your arm to move and place an object. Your brain sends signals though connective nerves to the synapses that trigger muscles. It also receives reports back that compare the command to the result-called a kinetic sense. Without looking, you always know where your arm is located and how fast it is moving. Then, if the item doesn't budge, you can add more power to get it going or at some point decide it's beyond your ability. If the object moves easily, you can back off on the force.

5.5 Direct Numerical Control

Direct numerical control (DNC) eliminates the need for tapes, disks, or drums to store information. The program is typed directly from a terminal to an NC machine tool or group of machines. If the machine tool is a CNC, the signal may run the machine directly or be stored in the memory of the machine computer.

Because NC machines do not have a computer, the prevailing view some years ago was that a central computer would supply the needed signals in "real time" directly to the NC machines while the machining operations were in process. It soon became apparent that if anything went wrong, all machines connected to the central computer would shut down. With the advent of CNC, it became possible to feed information from the large computer directly into the memory of each CNC computer. Now if the large central computer failed, the separate machine tools could rely on their own computers, and production could continue uninterrupted.

DNC has advantages when a great deal of control information is to be processed and used to operate many machine tools, or when the programs are lengthy and complex. It is also useful in flexible manufacturing situations where many machine tools are linked in a production line. The ability of a central computer to distribute information to many machine tools has caused the acronym DNC to mean "distributed numerical control".

5.6 Key Teams

1. Absolute Positioned System

A CNC drive system that retains its grid position even when turned off, then on. These machines feature linear feedback devices.

2. Ball-Screw

A backlash compensating, linear drive system in which a split nut is forced laterally both directions against a screw. Balls roll through the circular channel between nut and screw.

3. Closed Loop

A circuit that includes monitors to send a progress signal back to the CPU.

4. CPU (Central Processing Unit)

The component that decodes the program, sends the drive commands to the servo motors, and monitors and adjusts progress.

5. Feedback

The signal returning from the drive system to the CPU (Figure 5.2).

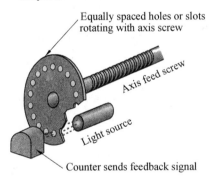

Figure 5.2 An indirect encoder counts pulses as the axis screw rotates

6. Hard Limits

Physical switches that verify location of an axis position.

7. Homing the Machine

Sending all axes to limit switches in order to achieve a zero base for the encoders. Generally required only on older machines.

8. Initialization

Turning on the control, depending on sophistication, routines must be followed to prepare for machining. Newer machines load software and ready themselves; older units need more preparation.

9. MCU
The master control unit see CPU.

10. Machine Home
A never-changing position where the machine must be driven to refresh its feedback signal at initial start-up.

11. Open Loop
A budget drive system that does not feature any provision for feedback signal processing.

12. Servo Error
An unwanted condition in a closed-loop system where the axis progress exceeds the tolerance from the expected result.

13. Servo Motor
A highly controllable motor with predictable power, speed, and acceleration curves based on input energy.

14. Soft limits
Optical sensors that verify proximity to hard limits switches.

15. Stepper Motor
A drive motor that moves a given part of a rotation, given one pulse of energy.

16. The Five Components
Using five electrical/mechanical components, a CNC drive performs the same tasks as using muscles to move and position your arm. The controller can keep a cutter hugging the true programmed shape from heavy cuts down to light loads, depending on machine construction and the nature of the shape and volume of metal being removed. The control constantly compares physical tool position to the expected position. If the cutter's progress doesn't fall within a preset lag tolerance, the controller has one or two options, depending on basic architecture within its operating system. The system performing these marvels is a closed-loop, linear, servo axis drive.

That mouthful is shortened to a CNC axis drive. It's responsible for the cutting motions and for the positioning accuracy. Like your muscle movements, five components are required (Table 5.1).

Table 5.1 Five muscle movements

CNC Drive	Muscle Motion
Controller unit	Brain-processing in/outgoing data
Axis drive relay card	Synapses-triggering muscles
Translation axis screw	Tendons pulling
Controllable drive motor	The muscle
Feedback device	Sensory nerves reporting progress

In order to achieve the maximum results and understand limitations, the action of each component should be generally understood by the machinist, which is our goal for this unit.

17. The Feedback Device

The fifth component is the watchdog in the system. Monitoring the axis progress and location, the feedback device is responsible for detecting and reporting position in real time back to the processor. There are two general families.

(1) Direct linear feedback.

Sometimes called absolute positioning, these systems make use of various kinds of calibrated distance sensors. The controller tracks axis position using signals from the scale. Direct feedback provides the advantage of acquiring position instantly during initial powering up. Machines equipped with direct feedback reacquire their axis and tool positions when the control is first turned on.

(2) Indirect feedback.

These devices monitor the amount of drive screw rotation called encoders (counters). About the only advantage of these encoding devices is their low cost compared to absolute positioning systems. The biggest disadvantage of encoders is that they can become lost either from depowering or from heavy machining loads. Machines of this type are set up such that they will not operate until they are homed after being powered up in the morning, so they have a zero base from which to begin positioning.

18. Initializing (Turning on a CNC Drive)

When homing the machine, or initialization (powering up the system) of encoding machines, each axis must be driven up to a reliable start point, usually against axis limit switches. That position is known as the machine home or axis home. Once parked at that never-changing hard-wired zero base location, usually at the far end of each axis, the controller then has a datum origin for its axis count.

There are anther kinds of linear and rotary feedback drive mechanisms, but scales and encoders represent most of the possibilities for common equipment. Still more advanced systems use reflected laser light as their axis sensing mechanism.

Words/Phrases and Expressions

CNC(computer numerical control) 数控系统
computer memory 计算机内存
miniaturization 小型化
axis drive 轴驱动
distributed numerical control 分布式数控
absolute positioned system 绝对定位系统
ball-screw 滚珠丝杠
open loop 开环
closed loop 闭环

CPU(central processing unit) 中央处理器
linear 线性的
backlash compensate 间隙补偿
monitor 监视器
feedback 反馈
decode 解码
encode 编码
limit 限位
switch 开关
home 归位
initialization 初始化
servo motor 伺服电机
stepper motor 步进电机

Chapter 6 Automatic Control

6.1 Open and Closed Loop Control

6.1.1 Introduction to Automatic Control

For our purpose of controlling machine tool, a control system may be defined as: one or more interconnected devices which work together to automatically maintain or alter the condition of the machine tool in a prescribed manner.

Such a system may be mechanical, electrical, electronic, hydraulic or pneumatic. In practice, many control systems are combinations of these and are termed hybrid systems.

In theory, an input signal is generatedin response to an inputted program command. This produces an output signal which turns a motor, which then moves the machine tool slide. In practice, however, to achieve this satisfactorily can be a complex problem.

One important distinction that must be made in relation to control systems, is between open loop and closed loop operation. Consider the following example.

6.1.2 Block Diagrams

By convention, control systems are represented on paper by block diagrams. This allows any system, regardless of power requirement, to be visualized simply and clearly. It is often known as the black box approach since a detailed knowledge of the workings of the component parts of the system is not required. It is only necessary to know how the output signal will respond to a given input signal and not what actually happens inside the box.

6.1.3 Feedback

The output quantity, is now having an effect on the input quantity. This system is now classified as a closed loop control system. To supply automatic control, the system is said to be error-actuated and, since the actual value is subtracted from the desired value (to determine the error), it is said to employ negative feedback. That is, we have an automatic closed loop control system employing negative feedback.

6.1.4 Closed Loop CNC Control

If this principle is now applied to CNC control, a program instruction becomes the command signal, an axis motor the controlled device, and the slide or axis position the controlled quantity. In physical terms, the command signal itself is unlikely to be large enough to "drive" an axis motor and would, in practice, have to be magnified by some sort of amplifier. The amount of this magnification is termed gain or, more correctly, loop gain and becomes very important in control system design. There will, in the case of a closed loop system, also need to be some means of monitoring slide position and some means of comparing input and output quantities, i. e. providing feedback. Figure 6.1 and Figure 6.2 show simplified block diagrams of typical open and closed loop control systems as applied to a controlled axis on a CNC machine tool.

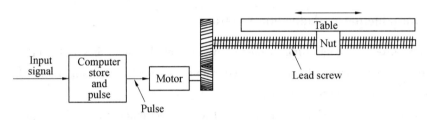

Figure 6.1   Block diagram of open loop control of machine tool axis

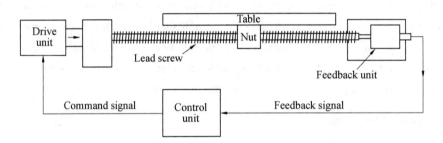

Figure 6.2   Block diagram of closed loop control of machine tool axis

Already we have built up a simplified model of a closed loop control system for a single axis of a CNC machine tool, and yet we do not require any detailed knowledge of the individual component parts that actually make the axis function. Such is the value of employing block diagrams in control system design.

The negative-feedback (error-actuated) closed loop concept has become the foundation for automatic control system design and is now widely applied in CNC control situations. The term servo-control is often used to describe such a system when applied to machine tool axis control. In fact, any fully automatic closed loop system, utilizing some form of magnification in which mechanical position is the controlled quantity, is known, more precisely, as a servomechanism.

We shall see too, however, that open loop control systems also have their part to play in CNC control applications.

Closed loop systems, of necessity, require more component parts, and extra control circuitry, in order for them to perform the feedback function, and they are consequently more complex than their open loop counterparts. This inevitably means higher costs for the design and implementation of closed loop systems on CNC machine tools. We would expect also that the performance (quality) of the machine tool would be correspondingly better to justify this increase in capital investment.

6.1.5 Automatic Adaptive Control

Adaptive control, by definition, is the process through which parameters are automatically optimized, or constrained, to achieve maximum production and dictated quality. As an example, feeds and speeds may be automatically adjusted to produce optimum tool performance, or feeds and speeds may be constrained to enhance surface finish requirements. Also, other factors may need to be either enhanced or constrained to produce dictated results; factors such as tool materials, workpiece materials, depth of cut, feeds and speeds, cutter life, and cutter geometry.

Automatic adaptive control (AAC) monitors, evaluates, and corrects performance as it happens. This process is still being refined. It will be the centerpiece of the automatic factory.

In an idealized block diagram of a closed loop that monitors an AAC process, the censors that measure performance accept data from the input and output end of the controller and from the output end of a designated process. It measures and evaluates this information and sends the evaluation to the controller through a decision-making section and a modification section. Accordingly, corrections are made in the process section.

This type of control could take place at a pallet station, a machine tool, an inspection or shipping station, and so on. In an automatic factory, it would take place at every phase through which the product passes. The goal is to control every facet of the manufacturing effort with computers. Computer control is to replace control by human beings.

6.2 Fundamental Problems of Control

There are a number of fundamental problems associated with designing control systems that are fully automatic. The way in which these problems are overcome will determine, to a great extent, the final performance or "quality" of the machine tool. This section examines some of the considerations associated with these problems.

6.2.1 Accuracy

It is appreciated in engineering that nothing can be perfectly accurate. This is demonstrated by the fact that most engineering components are produced with reference to both dimensional and geometrical tolerances. These are really statements of how inaccurately we are allowed to work. It is therefore more correct to say that something is accurate to within certain limits, where the limits are specified. A CNC control system is no exception. It shares these same limitations and is only accurate to within certain limits.

True accuracy will only ever be achieved by monitoring and controlling the position of the actual cutting tool edge relative to the established datums. Since this is impractical under present-day circumstances, discrete (and often ingenious) measuring devices have to be employed on machine tool elements that are relatively accessible.

The accuracy of the servo-control system therefore depends partly on the accuracy of the measuring device used to monitor the position of the machine slide, partly on how this measuring device is utilized, and partly on any mechanical inaccuracies present within the total system.

6.2.2 Resolution

The term resolution refers to the smallest increment, or dimension, that the control system can recognize and act upon. This is not the same as accuracy. For example, a measuring scale may have 20 divisions engraved upon it in which case its resolution would be 1 in 20. It may not be accurate, however, since the divisions could be unevenly spaced throughout its range.

When the range of a measuring device is large, overall resolution is sometimes obtained by utilizing a fine measuring device (over a small range) backed up by a coarse measuring device to permit the full range of slide motion to be measured. This principle is firmly established in the design of the common workshop micrometer.

6.2.3 Repeatability

As already stated, perfect accuracy is unattainable and so some dimensional tolerance must be applied. The component will be considered "correct" if its dimensions lie anywhere within this tolerance band. If a certain slide position is commanded many times in succession (as with many parts being made from the same part program), there will be a difference, or scatter, in the positions actually taken up by the slides. This scatter is a measure of the repeatability of the system. The repeatability of a system will always be better than its accuracy.

Indeed, it can be a considerable advantage if the same, rather than varying, machining errors appear on each repetition part. For example, the expensive processes of fitting and inspection could largely be eliminated.

6.2.4 Instability

Closed loop control tends to make for more accurate performance since the negative feedback is continually trying to reduce any error to an acceptable level. Under certain conditions, however, this continuous corrective action can lead to instability. Instability is the tendency for the system to oscillate about a desired position. It should be clear that instability is a function of closed loop control systems only, since it is entirely due to the characteristics of the feedback loop, the loop gain and the response of the system.

6.2.5 Response

The tuning of the response speed for any particular control system is a compromise between a minimal time lag and maintaining the stability of the system. It is a complex problem since to reach a commanded position from rest a slide first has to accelerate, achieve a steady state condition, and then decelerate onto the target with the minimum of overshoot. This has to be maintained under widely varying load conditions, i.e., under rapid traverse or feed, with an empty or fully loaded table.

6.2.6 Damping

To counter the effects of excessive overshoot or undershoot, and hence help minimise hunting, damping may be introduced into the system. A certain amount of damping will already be present within the system due to the effects of friction. Any undesirable oscillations will die out more quickly as the amount of damping increases. However, too much damping will cause the response of the system to be unnecessarily slow.

6.3 Types of Positional Control

One way of classifying CNC machine tools, other than by machine type, is by the positional control applied to their slide movements. This approach defines three broad headings which are sufficient to classify all CNC related equipment.

6.3.1 Positional or Point-to-point Control

Positional control is employed where the machine tool slides are required to reach a particular fixed coordinate point in the shortest possible time.

The path of getting from one point to another is not fixed or even important. In a typical tool path between programmed points on a point-to-point machine, at first it appears that the selected path between points is unpredictable and somewhat random.

6.3.2 Paraxial or Straight Line Control

Operations such as milling and turning, however, do dictate that the movement path between programmed coordinate points is important and, unlike drilling and boring operations, should be under full control at all times. Furthermore, speed control of the individual axes (commonly termed feed) must now also become a controlled variable in order for diagonal lines to be traced.

Straight line control is really a very limited form of contour control and to dedicate versatile computer control to such a menial task would be extremely wasteful. This system of control was more attractive on earlier NC machine tools since it was more versatile than point-to-point systems yet did not incur the increased costs of the more expensive continuous path systems. These controls utilized a method of moving from one point to another known as Linear Interpolation.

6.3.3 Continuous Path or Contouring Control

The method by which contouring systems move from one programmed point to another is called interpolation. This is a feature that merges the individual axis commands into a pre-defined tool path. There are three types of interpolation: linear, circular and parabolic. As stated previously most CNC controls now provide both linear and circular interpolation. Few controls used parabolic interpolation.

1. Linear Interpolation

This means moving from one point to another in a straight line. With this method of programming any straight line path can be traced. This will include all taper cuts. When programming linear moves, the coordinates must be given for the end of each line only, since the end of one line is the beginning of the next. The interpolator within the control unit calculates intermediate points and ensures that a direct path is traced by controlling and coordinating the speeds of the axis motors. Circles or arcs can only be programmed with some difficulty. The circle or arc must be broken down into a number of straight line moves. The smaller each of these segments becomes, the smoother the circle or arc will be.

2. Circular Interpolation

The programming of circles and arcs has been greatly simplified by the development of circular interpolation. Arcs up to 90° may be generated, and these may be further strung together to form half, three-quarter or full circles as desired.

Circular interpolation normally works from the current programmed position. The end point (X and Y coordinates) of the arc and the arc radius must also be specified. The circular interpolation control breaks the arc into small linear moves of high resolution. Circular interpolation is limited to one plane at a time. It is nonetheless a very versatile function and many free form shapes can be closely approximated to a series of arcs.

3. Parabolic Interpolation

The third method of interpolation, parabolic, is especially suited to mould and die manufacture where free form designs, rather than strictly defined shapes, are the norm. It easily adapts to those applications where aesthetic appeal is more important than mathematical description.

Parabolic interpolation positions the machine between three non-straight line positions in a movement that is either a part or a complete parabola. Its advantage lies in the way it can closely approximate curved sections with as much as 50 : 1 fewer points than with linear interpolation.

Such interpolation methods are ideally suited to the fast and accurate calculation abilities of the present-day digital computers. Most of the versatility of CNC machine tools is due to the undoubted flexibility resulting from such techniques. Certainly, programming components without them would be a long and arduous task.

Words/Phrases and Expressions

automatic control 自动控制
interconnect 相互连接
mechanical 机械的
electrical 电气的
electronic 电子的
hydraulic 液压的
pneumatic 气动的
response 响应
input signal 输入信号
output signal 输出信号
block diagram 框图
black box 黑匣子
error 误差
actual value 实际值
desired value 期望值
negative feedback 负反馈
amplifier 放大器
gain 增益
lead screw 丝杠
tool material 刀具材料
workpiece material 工件材料
depth of cut 切削深度
feed 进给
speed 速度
cutter life 刀具寿命
cutter geometry 刀具几何形状
accuracy 精度
resolution 分辨率
repeatability 可重复性
instability 不稳定性
oscillate 振荡
response 响应
empty load 空载
full load 满载

damping 阻尼
positional control 位置控制
point-to-point control 点对点控制
path 路径
continuous path 连续路径
contouring 轮廓
interpolation 插补
linear 线性
circular 圆弧
parabolic 抛物线

Chapter 7　Industrial Robots

7.1　Introduction

Robots are increasingly becoming a key feature of modern industrial manufacturing systems. But what is a robot and how does it differ from traditional "automation"? Automation can be described as the capability to operate without direct human intervention. Automatic devices have been on the industrial scene for more than 100 years. They are largely mechanical devices purpose-designed and built to perform a dedicated task. Because the configuration cannot easily be changed, a modern term labels this as hard automation. If the task changes, then this "hard" automation becomes redundant, or has to undergo physical modification to adapt its operation to suit the new task, or may be cannibalised and the components used again on other applications.

Industrial Robots also operate without direct human intervention and as such form a sub-class of automation. In addition, they exhibit certain other characteristics that set them apart from dedicated automatic devices.

1. Industrial Robots are Programmable

The movements and actions of an industrial robot can be determined initially by a human operator. The sequence of movements can be stored within a control system computer and the robot commanded to perform these movements repeatedly and with great accuracy. If the task changes, the robot can be "taught" a new sequence of movements thus easily adapting to the new task. Since operation is largely under software control (i.e. the control of a computer program), robots can be thought of as "soft" automation.

2. Industrial Robots Have Multiple Movements

Industrial robots can perform three-dimensional manoeuvres in space. Six degrees of freedom of manipulation are provided on most industrial robots, which make them adaptable to a wide variety of industrial tasks. Additional degrees of freedom may be provided to increase the accessibility of the robot for certain applications. The three-dimensional space within which the robot can reach is known as its working envelope.

3. Industrial Robots Have Interchangeable Grippers

Most industrial tasks involve transporting or manipulating tools, components, appliances or applicators to perform various manufacturing activities. The majority of these tasks involve the holding or gripping of items of varying size and shape. Since it is almost impossible to design a universal holding device, and in order to retain the flexibility of robot applications, most robots can be fitted with a variety of "hands". These are known as end effectors and will often be purpose-designed for a particular task. They may be mechanical, electro-magnetic, or pneumatic (vacuum cup) devices or consist simply of hooks or holders. It is becoming more common for robots to be applied to more than one task simultaneously, thus necessitating an end effector change from one task to another. This is similar to the automatic tool changing facility available on many CNC machine tools.

On the basis of the above characteristics, an industrial robot may be defined as: a computer-controlled, re-programmable mechanical manipulator with several degrees of freedom capable of being programmed to carry out more than one industrial task.

7.2 Robot Configurations

Industrial robots can perform manoeuvres within a three-dimensional working envelope and the extent of this working envelope will be dictated by the configuration, or geometry, of the robot employed. Four robot configurations can be identified.

1. Cartesian

This configuration is illustrated in Figure 7.1. It may also be referred to as a rectangular configuration. The three perpendicular axes will generate a three-dimensional rectangular-shaped working envelope.

2. Cylindrical

This configuration is illustrated in Figure 7.2. The geometry combines vertical and horizontal linear movement with rotary movement in a horizontal plane. As its name implies, a cylindrical working envelope will be generated by this configuration.

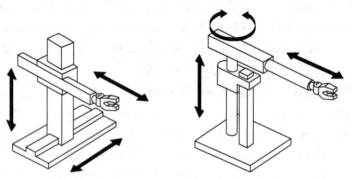

Figure 7.1　Cartesian　　　　Figure 7.2　Cylindrical

3. Polar

The polar coordinate robot, sometimes called spherical coordinate robot. This configuration is illustrated in Figure 7.3. It has three axes of motion: two of them are rotary, and the third is linear. It may also be referred to as a spherical configuration. Rotational movement in both horizontal and vertical planes is combined with a single linear movement of the arm. The working envelope will be largely spherical but with a conical-shaped "dead zone" above the base due to the inability of the geometry to "reach" this area.

4. Articulated

This configuration is illustrated in Figure 7.4. It may also be referred to as an angular configuration. Jointed movements are provided that approximate to the joints of the human body. Terms such as waist, shoulder, elbow and wrist are used to describe the movements and indicate which joint is operative. The working envelope will be spherical.

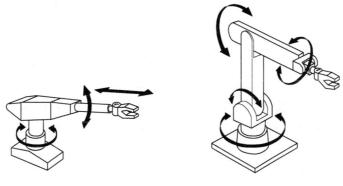

Figure 7.3   Polar   Figure 7.4   Articulated

Whilst the above configurations imply only three degrees of freedom, the full six degrees of freedom are usually accommodated by an end effector possessing the three "missing" degrees of freedom.

All types of industrial robot need to be programmed. There are predominantly three modes by which this may be accomplished.

5. Walk-Through Programming

In this mode, the robot is moved by remote control to the various positions it must visit. This is commonly done via a key-pad or teach pendant. At each position, a "teach" button is pressed to retain the position in the control system memory.

6. Lead-Through Programming

Often referred to as "lead by the nose" programming. This system involves the operator physically guiding the robot through the desired path and sequence of events. The control system stores the entire movement sequence and reproduces it faithfully. This mode is ideally suited to applications such as paint spraying where the experience and dexterity of the human operator need to be reproduced.

7. Off-Line Programming

This mode can be used to program the movements and actions of the robot without itnecessarily being caused to traverse the desired path. It is also often used as a convenient means of editing or modifying existing sequences. It is usually accomplished via a computer using a specialised robot programming language.

7.3 Robot Generations

Robot technology, like electronic technology, is developing in distinct phases or generations. At present, three generations can be identified.

First-generation robots can be likened to devices operating under open loop control. They invariably work through a preprogrammed sequence of operations whether work is present or not. They cannot detect any change in the surrounding environment and cannot therefore modify their actions accordingly.

Second-generation robots are robots equipped with a range of sensors, and the necessary computing power, to modify their actions due to small changes in the surrounding environment. For example, proximity (touch) sensors can detect the presence or absence of components and initiate the robot cycle accordingly. Vision sensors can differentiate between a number of components and "instruct" the robot to execute a different sequence of events depending on the component identified.

Third-generation robots, at present only in the research stage, will be characterized by their ability to plan, make decisions, and execute tasks "intelligently". They are likely to be programmed to operate so as to maximise some defined objective.

First-generation robots form the great majority of industrial robots in current use. Second-generation robots are under continual development and used in limited numbers. Third-generation robots are still in the research stage. Development of third-generation robots will depend to a large extent on parallel developments being made in Artificial Intelligence (AI) software systems.

7.4 Industrial Applications of Robots

The industrial applications of robots are many. Certain factors need to be considered before applying robots to aspecific task. These include: reach, working envelope, lifting capacity, gripping capacity, accuracy, manoeuvrability, accessibility, repeatability, speed, form of motion (point-to-point vs. continuous path), and the provision of sensory feedback. The more common applications include followings.

(1) Loading and unloading of manufacturing processes: press tools, injection molding and die casting machines, conveyors, furnaces, etc..

(2) Palletising and packaging processes: packing and loading of components onto pallets.

(3) Welding and flame cutting processes: seam, spot and continuous run.

(4) Coating processes: dip and spray applications.

(5) Fettling and cleaning processes: steaming, blasting, grinding, polishing, deburring.

(6) Assembly processes: pick and place, store and retrieve.

(7) Applicating and dispensing processes: pouring, scooping, ladling-adhesive, mastic, liquid, powder.

(8) Testing and manoeuvring processes: X-ray, radioactive environments.

Robots are employed in industrial applications for one or more of the following reasons.

(1) They can carry out boring and repetitive tasks with accuracy and consistency, thus freeing human operators from mindless tasks.

(2) They can replace human operators in dangerous, hostile or obnoxious environments, thus eliminating unhealthy working conditions.

(3) They can achieve flexible operation and savings in manpower by being able to operate at unsocial hours and requiring minimum services resulting in significant cost savings.

(4) They are stronger and have greater reach than human operators.

(5) They can achieve consistent, predictable quality and productivity for sustained periods of time.

(6) They are cheaper to employ and run than human operators.

(7) They offer a low-cost route into Flexible Manufacturing.

Words/Phrases and Expressions

robot 机器人
cannibalise 拆用(旧零件修理或装配另一部机器),同类相食(以推出一种类似新产品)
manoeuvre 动作,操纵
programmable 可编程的
freedom 自由度
end effector 末端执行器
electro-magnetic 电磁的
configuration 配置,结构,构造
cartesian 笛卡儿的
cylindrical 圆柱形的
polar 极坐标
spherical 球坐标的
articulated 关节的
dead zone 死区
off-line programming 离线编程
Artificial Intelligence (AI) 人工智能
load and unload 装卸
press tool 冲床
injection molding 注塑成型
die casting machine 压铸机
conveyor 输送机

furnaces 熔炉
palletising 码垛
packaging 包装
pallet 托盘
welding 焊接
flame cutting 火焰切割
seam 接缝
spot 点焊
coating process 涂层工艺
dip 浸涂
spray 喷涂
steam 蒸
blast 喷砂
grind 研磨
polish 抛光
debur 毛刺
assemble 组装
pour 浇注
X-ray X射线
radioactive 放射性的
flexible manufacture 柔性制造

Chapter 8　Nontraditional Machining Operations

The need for nontraditional machining processes came actually as a result of the shortcomings and limitations of the conventional, mechanical, chip-generating processes. While conventional processes can be applied only to soft and medium-hard materials, very fine features of extremely hard materials can be produced using the nontraditional machining processes. The latter include a variety of processes, each has its advantages and field of applications. Following is a brief discussion on each of the nontraditional processes commonly used in industry.

8.1　Ultrasonic Machining

Ultrasonic machining is particularly suitable for machining hard, brittle materials because the machining tool does not come in contact with the workpiece. They are separated by a liquid (vehicle), in which abrasive grains are suspended. Equal volumes of water and very fine grains of boron oxide are mixed together to produce the desired suspension. Ultrasonic energy, which is applied to the tool, results in high-frequency mechanical vibrations (20 to 30 kHz). These vibrations impart kinetic energy to the abrasive grains, which, in turn, impact the workpiece and abrade its material. The machining tool must be made of a tough ductile material such as copper, brass, or low-carbon steel, so that it will not be liable to fretting wear or abrasion, as it is the case with the workpiece.

Ultrasonic machining is employed mainly in making holes with irregular cross sections. Both through and blind holes can be produced by this method.

8.2　Abrasive-Jet Machining

Liquids, in which abrasive particles are suspended, are pumped under extremely high pressure out from a nozzle. The resulting jet stream is then employed in processes like deburring, drilling, and cutting of thin sheets and sections. The process is particularly advantageous when cutting glass and sheets of composites. The shortcomings of this process involve the problems associated with using high-pressure pumps and the relatively slow feed rate employed.

8.3　Chemical Machining

Chemical machining involves attacking the surfaces of the workpiece to be machined by an etchant, which reacts with the metal and dissolves the resulting chemical compound. The procedure consists of first covering the surfaces of the workpiece that are not to be machined with neoprene rubber or enamel and then dipping the workpiece for a while into a basin full of all appropriate etchant. Very fine details can be etched by this method, and the quality of the machined surface is high and free from any chips. A further advantage of this process is that it does not result in any work hardening.

8.4 Electrochemical Machining

The mechanism with which electrochemical machining (ECM) takes place is reciprocal to that of the electroplating process, although similar equipment is used in both cases. In electrochemical machining, the workpiece is connected to the anode, while the cathode is connected to a copper ring that is used as the machining tool. Low-voltage, high-amperage direct current is used, and an electrolyte is pumped into the small gap between the workpiece and the copper ring. As was the case in electroplating, the amperage plays an important role in determining the rate of metal transfer from the anode to the cathode, i.e., the rate of metal removal during the electrochemical machining process.

Electrochemical machining can be applied to all electrically conductive metals, including hardened alloy steel and tungsten, and is particularly advantageous for machining thin sheets of nickel and titanium.

8.5 Electrodischarge Machining

Electrodischarge machining(EDM) is used for producing parts having intricate shapes and can be applied to all metallic materials, whether they are ductile or brittle. Nevertheless, it cannot be used with ceramics, plastics, or glass. Metal removal takes place as a result of an electric arc between the electrode and the workpiece, which are kept apart. A dielectric liquid like kerosene is pumped through the small gap of about 0.02 in(0.5 mm) between the electrode and the workpiece, as shown in Figure 8.1. The dielectric fluid also acts as a coolant and a flushing medium to whip away the removed metallic dust.

The electrode is usually made of a material that can easily be shaped, such as copper, brass, graphite, copper-tungsten mixture, and the like. The electrode must be given a shape that fits exactly into the desired final cavity. Consequently, intricately shaped parts can easily be produced by this method, which has gained widespread industrial application in the manufacture of tools, metal-forming and forging dies, plastic, and die-casting molds. Generally, it can be stated that the quality of the machined surface is dependent upon the number of sparks (electric arc sparks) per second, which range from 3 000 up to 10 000.

A new version of EDM is wherein the conventional electrode is replaced by a tensioned wire of copper or tungsten, which is guided by a CNC system to trace any desired contour. In

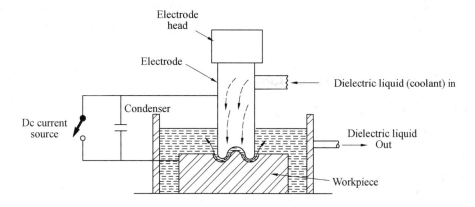

Figure 8.1 A schematic diagram of the EDM process

fact, this process has revolutionized the tool and die-making industry. While sharp corners are avoided in tools manufactured by conventional processes to avoid breakage or cracking during subsequent heat treatment, wire EDM can cut heat-treated steels directly to the desired shape. Therefore, large dies having intricate shapes and sharp corners can be produced by this technique.

8.6 Chem-Milling

In the chem-milling (trade name) process, the material is removed by dissolution rather than cutting action. Three steps are involved in the process. First, the work is masked with a paint or vinyl-type plastic in the areas where no metal removal is desired. Silk screening is often used to apply the paint, the part is then immersed in an etching fluid, which may be acidic or basic, and it is allowed to remain until the required depth has been obtained. Lastly, the maskant is removed from the work. This process was originally developed for alloys of magnesium and aluminum, but is now being extended to other materials. Dimensional control can be quite accurate.

8.7 Use of Lasers

The principle of the laser is the emission of energy at a predetermined level of excitation. Since there does not exist an infinite number of excitation states for any given material, it is necessary for an electron to change its state from one level to a higher one by absorbing energy and thereby becoming excited. The laser beam is a very concentrated monochromatic beam of extremely high intensity.

The application of lasers offers advantages over certain other processes. For example, contact between the laser and the workpiece is not required. The high-power densities available make it possible to vaporize any material, and the small spot permits the removal operation to be performed in microscopic regions. Laser beams easily drill holes in diamond dies for wire drawing.

Words/Phrases and Expressions

nontraditional machining process 特种加工
ultrasonic machining 超声波加工
abrasive grain 磨粒
suspend 悬浮
boron oxide 氧化硼
copper 铜
brass 黄铜
low-carbon steel 低碳钢
abrasive-jet machining 磨料喷射加工
pump 泵
nozzle 喷嘴
sheet of composite 复合板材
chemical machining 化学加工
etchant 腐蚀剂
neoprene rubber 氯丁橡胶
enamel 搪瓷
chip 切屑
electrochemical machining 电化学加工
reciprocal 相反的
electroplate 电镀
anode 阳极
cathode 阴极
amperage 安培
electrolyte 电解液
gap 间隙
electrodischarge machining 放电加工
hardened alloy steel 硬质合金钢
tungsten 钨
ductile 韧性
brittle 脆性
ceramics 陶瓷
plastics 塑料
glass 玻璃

electric arc 电弧
electrode 电极
dielectric liquid 电解液
kerosene 煤油
form 成型
forge 锻造
chem-milling 化学铣削
vinyl-type plastic 乙烯塑料
maskant 保护层
magnesium 镁
laser 激光
vaporize 蒸发
laser beam 激光束
diamond 金刚石
die 模具

Chapter 9　Rapid Prototyping and Manufacturing

9.1　Introduction

To substantially shorten the time for developing patterns, moulds, and prototypes, some manufacturing enterprises have started to use rapid prototyping (RP) methods for complex patterns making and component prototyping. Over the past few years, a variety of new rapid manufacturing technologies, generally called Rapid Prototyping and Manufacturing (RP&M), have emerged; the technologies developed include Stereolithography(SL), Selective Laser Sintering (SLS), Fused Deposition Modeling(FDM), Laminated Object Manufacturing(LOM), and Three-dimensional Printing(3D Printing). These technologies are capable of directly generating physical objects from CAD databases. They have a common important feature: the prototype part is produced by adding materials rather than removing materials, that is, a part is first modeled by a geometric modeler such as a solid modeler and then is mathematically sectioned (sliced) into a series of parallel cross-section pieces. For each piece, the curing or binding paths are generated. These curing or binding paths are directly used to instruct the machine for producing the part by solidifying or binding a line of material. After a layer is built, a new layer is built on the previous one in the same way. Thus, the model is built layer by layer from the bottom to top.

9.2　Rapid Prototyping and Manufacturing Technologies

As mentioned earlier, there are several technologies available for model production based on the principle of "growing" or "additive" manufacturing. The major differences among these technologies are in two aspects: materials used; part building techniques. The following sections will explain in detail these rapid prototyping technologies with respect to the above two aspects.

9.2.1　Stereolithography(SLA)

Stereolithography apparatus was invented by Charle Hull of 3D Systems Inc. It is the first commercially available rapid prototyper and is considered as the most widely used prototyping machine. The material used is liquid photo-curable resin, acrylate. Under the initiation of photons, small molecules (monomers) are polymerized into large molecules. Based on this principle, the part is built in a vat of liquid resin as shown in Figure 9.1. The SLA machine creates the prototype by tracing layer cross-sections on the surface of the liquid photopolymer pool with a laser beam. Unlike the contouring or zigzag cutter movement used in CNC machining, the beam traces in parallel lines, or vectorizing first in one direction and then in the orthogonal direction. An elevator table in the resin vat rests just below the liquid surface whose depth is the light absorption limit. The laser beam is deflected horizontally in X and Y axes by galvanometer-driven mirrors so that it moves across the surface of the resin to produce a solid pattern. After a layer is built, the elevator drops a user-specified distance and a new coating of liquid resin covers the solidified layer. A wiper helps spread the viscous polymer over for building the next layer.

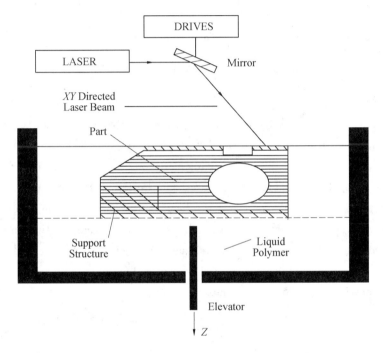

Figure 9.1 Stereolithography

9.2.2 Selective Laser Sintering(SLS)

DTM Corp. (Austin TX) offers an alternative to liquid-curing systems with its selective laser sintering (Figure 9.2) systems which were developed by Carl Deckard and Joseph Beaman at the Mechanical Engineering Department of University of Texas at Austin. SLS uses a carbon dioxide laser to sinter successive layers of powder instead of liquid. In SLS processes, a thin layer of powder is applied by a counter-rotating roller mechanism onto the work place. The powder material is preheated to a temperature slightly below its melting point. The laser beam traces the cross-section on the powder surface to heat up the powder to the sintering temperature so that the powder scanned by the laser is bonded. The powder that is not scanned by the laser will remain in place to serve as the support to the next layer of powder, which aids in reducing distortion. When a layer of the cross-section is completed, the roller levels another layer of powder over the sintered one for the next pass.

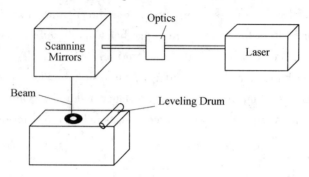

Figure 9.2　Working principle of SLS

9.2.3　Fused Deposition Modelling(FDM)

Another rapid prototyping system, developed by Stratasys Inc., constructs parts based on deposition of extruded thermoplastic materials, is called FDM process (Figure 9.3). The spool of thermoplastic filament feeds into a heated FDM extrusion head. The movement of the FDM head is controlled by computer. Inside the flying extrusion head, the filament is melted into liquid (1° above the melting temperature) by a resistant heater. The head traces an exact outline of each cross-section layer of the part. As the head moves horizontally in X and Y axes, the thermoplastic material is extruded out of a nozzle by a precision pump. The material solidifies in 1/10 second as it is directed onto the workplace. After one layer is finished, the extrusion head moves up a programmed distance in Z direction for building the next layer. Each layer is bonded to the previous layer through thermal heating.

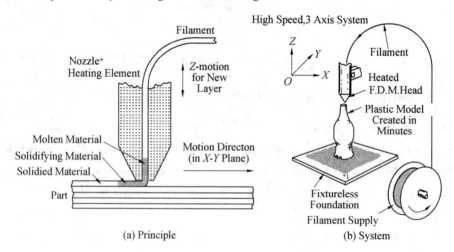

Figure 9.3　Fused Deposition Modeling

9.2.4 Laminated Object Manufacturing (LOM)

The LOM processes produce parts from bonded paper plastic, metal or composite sheet stock (Figure 9.4). LOM machines bond a layer of sheet material to a stack of previously formed laminations, and then a laser beam follows the contour of the part cross-section generated by CAD to cut it to the required shape. The layers can be glued or welded together and the excess material of every sheet is either removed by vacuum section or remains as next layer's support.

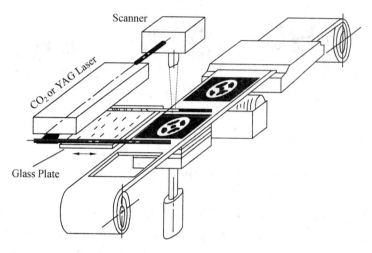

Figure 9.4 Laminated Object Manufacturing

9.2.5 Three-Dimensional Printing (3D Printing)

Three-dimensional printing was developed at Massachusetts Institute of Technology. In the 3D Printing process, a 3D model is sliced into 2D cross-section layers in computer. A layer of powder is spread on the top of the piston, the powder bed, in a cylinder, and then an inkjet printing head projects droplets of binder material onto the powder at the place where the solidification is required according to the information from the computer model. After one layer is completed, the piston drops a predefined distance and a new layer of powder is spread out and selectively glued. When the whole part is completed, heat treatment is required to enhance the bonding of the glued powder, and then the unbonded powder is removed (Figure 9.5).

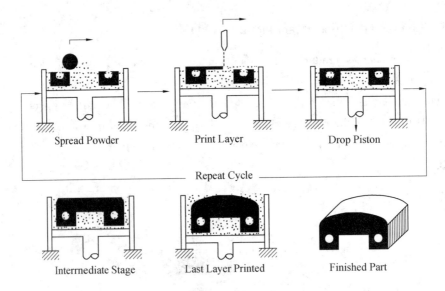

Figure 9.5　3D Printing

9.3　Current Application Areas of RP&M

During the successful application in various areas in recent years, RP is being used as a communication and inspection tool in the procedure of product development and realization of the rapid feedback of the design information. Although RP&M technologies are still at their early stage, a large number of industrial companies have benefited from applying the technologies to improve their product development. This kind of dynamic, controllable and simultaneous product development system should be realized under the structure.

Many of the unit technologies in the system are already mature. But some techniques should be selected and how to integrate those technologies effectively remain difficult problems to solve.

9.3.1　Direct Manufacturing of Metal Parts with RP

RP technology has proven successful in many ways as a process which can be easily and rapidly automated and with almost no geometry limitation in parts to be fabricated. But limitations of this kind of technology are mainly them availability of materials. Commonly materials used in present RP systems are polymers, paper and ceramic. Metal parts directly built with RP are comparatively rare. It is obvious that with metal, the technology would be extended to more useful prototypes or parts and fast manufacturing of tools. The most important requirements RP should fulfill are to build metal parts with adequate strength and accuracy. That is the main problem existing now.

9.3.2 Combination of RP and Metal Casting, Rapid Tooling

Although direct manufacturing of metal parts with RP is not well developed, indirect methods have been found and shown feasible through the combination of RP and metal casting (e. g. ,investment casting). The major impact RP technology has its ability to fabricate high quality and complex patterns used in investment casting with lower cost and shorter leading times. Although application in mass production is not practical, this kind of technology is very suitable for one case or low volume production. RT (rapid tooling) is one of the best suitable application areas. Many case studies with different RP processes and casting techniques have proven to be successful.

Table 9.1 shows the most common commercial RP systems and the materials of prototypes produced by them. The application approaches in casting or tooling with different kinds of prototypes are listed in Table 9.2.

Table 9.1 RP Processes with materials

RP Processes	Materials
SLA	Polymer
LOM	Paper
SLS	Ceramic, wax and alloy
FDM	Wax and polymer
3DP	Ceramic and wax

Table 9.2 Conversion approaches in casting or tooling

Kind of Material	Conversion Approach
Polymer	Quick casting, epoxy pattern, soft tooling
Paper	Ceramic investment casting, sand casting
Ceramic	Direct ceramic casting mold
Wax	Lost wax casting
Alloy	Soft tooling
Arbitrary	Metal spray, EDM

According to the ability of RP&M and casting techniques, feasible process combinations can be chosen. During the scheme of process selection, the reliable criteria should be followed. Factors including machining time, cost, part quality and environment effect need to be considered.

9.4 Rapid Prototyping and Manufacturing Problems

There are still some problems in RP&M. The physical models made by most of these systems can not be used as working parts, mostly due to material and economic constraints. The major problems in the current RP&M systems include: part accuracy, limited material variety and mechanical performance.

9.4.1 The Problem about Part Accuracy

A large number of factors limit the ability of rapid prototyping systems to create parts as accurate as the CAD designs on which they are based. The most common sources of error among the RP&M systems could be categorized as mathematical, process-related or material-related.

(1) Mathematical errors include facet approximation of the part surfaces in the standard input to RP&M systems.

(2) Process-related errors affect the shape of the layer in the X-Y plane and along the Z axis, the registration between different layers and the overall 3D shape. These errors are mainly dependent on the accuracy of the RP&M machines and operators' experiences.

(3) Major material related errors are shrinkage and distortion.

9.4.2 The Problem about materials

The current RP&M systems use very limited types of materials. Parts built by RP&M systems tend to be weak and fragile compared to those made conventionally from metals and engineering plastics. Some materials for the RP&M machines are expensive or toxic. Most of the research and development efforts have been focused on improving part materials, which are carried out in two different directions.

One is that plastic companies with products already on the market, particularly those based on SL, SLS and FDM, have developed plastics with better physical properties-less brittle, lower shrink, lower viscous resin and similar to the end-user applications.

The other direction is to focus on metals. The users of RP&M technologies tend to build models with materials whose properties are similar to the materials they might use in their end-use applications. Metal is most-commonly used in current industry. All rapid prototyping machines on the market can be used to produce metal parts indirectly through various casting processes such as investment casting. The direct production of metal parts is still in development.

Words/Phrases and Expressions

Rapid Prototyping (RP) 快速成型
Stereolithography(SL) 立体光刻
Selective Laser Sintering(SLS) 选择性激光烧结
Fused Deposition Modeling(FDM) 熔融沉积建模
Laminated Object Manufacturing(LOM) 分层实体制造
Three-dimensional Printing(3D Printing) 3D 打印
Photo-Curable Resin 光固化树脂
acrylate 丙烯酸酯

polymerize 聚合
photopolymer 光敏聚合物
resin 树脂
solidify 固化
carbon dioxide 二氧化碳
counter-rotate 反向旋转
melting point 熔点
scan 扫描
wax 蜡
epoxy 环氧树脂

Chapter 10 Computer Integrated Manufacturing

10.1 Introduction

The term computer integrated manufacturing (CIM) is often used interchangeably with CAD/CAM. However, CIM has a slight broader meaning than CAD/CAM. The CAD/CAM functions are dealing with all the aspects of design and manufacturing and all the production processes involved such as material planning, production scheduling etc. and CAD/CAM establishes the direct link between the design and manufacturing processes by automating all steps leading from the one function to the other.

Computer integrated manufacturing though includes all the engineering activities included in the CAD/CAM and includes all the business functions as well. The ideal CIM system applies computer technology to all of the operational functions and information processing functions in manufacturing from order receipt, through design and production, to product shipment.

The computer presence is throughout the firm, touching all activities that support manufacturing. In this integrated computer system, the output of each activity serves as the input to the next. Thus a chain of activities is interlinked and begins with sales orders and ends with shipment of the product.

Although CIM encompasses many of the other advanced manufacturing technologies such as computer numerical control (CNC), computer-aided design/computer-aided manufacturing (CAD/CAM), robotics, and just-in-time delivery (JIT), it is more than a new technology or a new concept.

Computer integrated manufacturing (CIM) is the term used to describe the modern approach to manufacturing. Computer integrated manufacturing is an entirely new approach to manufacturing, a new way of doing business. To understand CIM, it is necessary to begin with a comparison of modern and traditional manufacturing. Modern manufacturing encompasses all of the activities and processes necessary to convert raw materials into finished products, deliver them to the market, and support them in the field. These activities include the following (Figure 10.1):

(1) Identifying a need for a product.

(2) Designing a product to meet the needs.

(3) Obtaining the raw materials needed to produce the product.

(4) Applying appropriate processes to transform the raw materials into finished products.

(5) Transporting products to the market.

(6) Maintaining the product to ensure proper performance in the field.

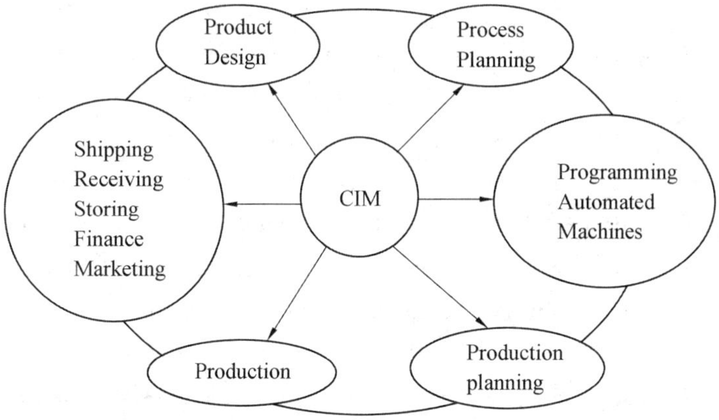

Figure 10.1　Major components of CIM

10.2　Historical Development of CIM

Material requirements planning (MRP) is an important concept with a direct relationship to CIM. It is a process that can be used to calculate the amount of raw materials that must be obtained in order to manufacture a specified lot of a certain product. It has taken many years for CIM to develop as a concept, but integrated manufacturing is not really new. In fact, integration is where manufacturing actually began. Manufacturing has evolved through four distinct stages:

1. Manual Manufacturing

Manual manufacturing using simple hand tools was actually integrated manufacturing. All information needed to design, produce, and deliver a product was readily available because it resided in the mind of the person who performed all of the necessary tasks. The tool of integration in the earliest years of manufacturing was the human mind of the craftsman who designed, produced, and delivered the product.

2. Mechanization/Specialization

With the advent of the industrial revolution, manufacturing processes became both specialized and mechanized. Instead of one person designing, producing, and delivering a product, workers and/or machines performed specialized tasks within each of these broad areas. Communication among these separate entities was achieved using drawings, specifications, job orders, process plans, and a variety of other communication aids. To ensure that the finished product could match the planned product, the concept of quality control was introduced. The disadvantage is that the lack of integration led to a great deal of waste.

3. Automation

Automation improved the performance and enhanced the capabilities of both people and machines within specialized manufacturing components. For example, CAD enhanced the capability of designers and drafters. CNC enhanced the capabilities of machinists and computer assisted planners. But the improvements brought on by automation were isolated within individual components or islands. Because of this, automation did not always live up to its potential. However, if these various automated subsystems were not tied together in a way that allowed them to communicate and share accurate, up-to-date information instantly and continually. The same limitations apply in an automated manufacturing setting. These limitations are what led to the current stage in the development of manufacturing, integration.

4. Integration

With the advent of the computer age, manufacturing has developed full circle. It began as a totally integrated concept and, with CIM, has once again become one. However, there are major differences in the manufacturing integration of today and that of the manual era of the past. First, the instrument of integration in the manual era was the human mind. The instrument of integration in modern manufacturing is the computer. Second, processes in the modern manufacturing setting are still specialized and automated. The components as design, planning, and production have evolved both in processes and in the tools and equipment used to accomplish the processes. These individual components of manufacturing evolved over the years into separate islands of automation. However, communication among these islands was still handled manually. This limited the level of improvement in productivity that could be accomplished in the overall manufacturing process. When these islands and other automated components of manufacturing are linked together through computer networks, these limitations can be overcome.

5. Internet

"Networking" is the only way for the development of modern integrated manufacturing technology. The manufacturing industry is becoming integrated and orderly, which is synchronized with the development of human society. In the market competition of the manufacturing industry, there are many pressures: increasing costs, accelerating product updates, changing market demands, rapid development of customer order production methods, increasing impact of global manufacturing, etc. ; companies must be in production organizations implement some kind of profound change. The development of science and technology, especially computer technology and network technology, has made this change possible.

6. Intelligent

Intelligent manufacturing technology is the prospect of manufacturing technology development. The basis of the intelligent manufacturing model is the intelligent manufacturing system. The intelligent manufacturing system is not only the application environment formed by the integration of intelligence and technology, but also the carrier of the intelligent manufacturing model. Compared with traditional manufacturing, the intelligent manufacturing system has the following characteristics: human-machine integration; self-discipline ability; self-organization and ultra-flexibility; learning ability and self-maintenance ability; in the future, it has more advanced human-like thinking ability.

7. Green Manufacturing

"Green" is quoted from the field of environmental protection. The development of human society is bound to move towards harmony between human society and nature, including manufacturing technology.

Manufacturing products must take environmental protection into full consideration from the design stage to the design stage, manufacturing stage, sales stage, use and maintenance stage, to the recycling stage and remanufacturing stage. As a "green" manufacturing, products must also be works of art to a certain extent, to adapt to the user's production, work, and living environment, to give people a noble spiritual enjoyment, reflecting the height of material civilization, spiritual civilization and environmental civilization blend.

8. Standardization of CIMS

In the process of manufacturing industry's development towards globalization, networking, integration and intelligence, standardization technology has become more and more important. It is the foundation of information integration, function integration, process integration and enterprise integration.

10.3 The CIM Wheel and Benefits

This broad, modern view of manufacturing can be compared with the more limited traditional view that focused almost entirely on the conversion processes. The old approach excluded such critical pre-conversion elements as market analysis research, development, and design, as well as such after-conversion elements as product delivery and product maintenance. In other words, in the old approach to manufacturing, only those processes that took place on the shop floor were considered manufacturing. This traditional approach of separating the overall concept into numerous stand-alone specialized elements was not fundamentally changed with the advent of automation. With CIM, not only are the various elements automated, but the islands of automation are all linked together or integrated. Integration means that a system can provide complete and instantaneous sharing of information (Figure 10.2).

1. The CIM System

The Computer and Automated Systems Association (CASA) of the Society of Manufacturing Engineer (SME) developed the CIMsystem as a way to comprehensively but concisely illustrate the concept of CIM. The CASA/SME developed the CIM system to include several distinct components:

(1) Manufacturing management/human resource management.

(2) Marketing.

(3) Strategic planning.
(4) Finance.
(5) Product/process design and planning.
(6) Manufacturing planning and control.
(7) Factory automation.

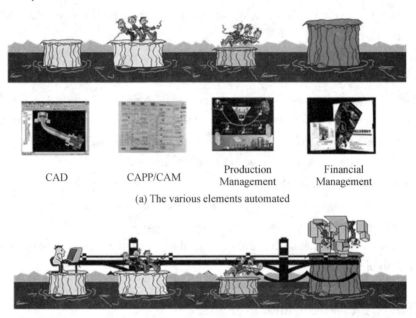

(c) Integrated means that a system can provide complete and instantaneous sharing of information. Integration is accomplished by computers

Figure 10.2　Islands of automation

2. Benefits of CIM

Fully integrated manufacturing firms realize a number of benefits from CIM:
(1) Product quality increases.
(2) Lead times are reduced.
(3) Direct labor costs are reduced.
(4) Product development times are reduced.
(5) Inventories are reduced.
(6) Overall productivity increases.

(7) Design quality increases.

10.4 Manufacturing Resources Planning and CIM

Material requirements planning (MRP) is an important concept with a direct relationship to CIM. It is a process that can be used to calculate the amount of raw materials that must be obtained in order to manufacture a specified lot of a certain product. Material requirements planning involves using the bill of material, production schedule, and inventory record to produce a comprehensive, detailed schedule of the raw materials and components needed for a job.

As manufacturing technology has evolved from automation to integration, MRP has also evolved. The acronym MRP now means manufacturing resources planning. This broader concept goes beyond determining material requirements to also encompass financial tracking and accounting.

MRP is an information system that integrates all manufacturing and related applications, including decision support, material requirements planning (MRP), accounting and distribution.

ERP is short for enterprise resource planning, a business management system that integrates all facets of the business, including planning, manufacturing, sales, and marketing.

The modern version of MRP is particularly well suited to the integrated approach represented by CIM. In this approach, MRP can be an effective inventory planning and control tool.

Key concepts relating to MRP include:

(1) Independent and dependent demand.

(2) Lead times.

(3) Common-use items.

Independent demand resources are those that are not tied to any other resource. They stand alone.

Dependent demand resources are tied directly to other resources. Receiving a dependent resource without the other resources it requires does no good. In manufacturing, resources are more likely to be dependent than independent. Raw materials, in-progress parts, components, and subassemblies are all part of the overall manufacturing inventory.

Lead times are of two types: ordering lead time and manufacturing lead time. Ordering lead time is the total amount of time between initiating a purchase order and receiving the goods. Manufacturing lead time is the total amount of time required to perform all the preparative steps necessary to produce a given part. Lead times are important because they are used in developing the schedules for ordering materials and producing products. They are also where MRP is most likely to break down. Resource planners depend on the lead times provided to them by other personnel. If these times are padded or inaccurate, the MRP results will be equally inaccurate. Common-use items are items used in producing more than just one product.

Manufacturing resources planning results in a variety of products of value to manufacturing managers in addition to the master schedule:

(1) Release notices that notify the purchasing department to place orders.

(2) Revise schedules showing updated due dates.

(3) Cancellation notices that notify appropriate personnel of cancellations that result from changes to the master schedule.

(4) Inventory status reports.

Manufacturing resources planning is the most appropriate planning approach for a CIM setting. When completely implemented it can result in a number of benefits:

(1) Inventory reduction.

(2) Quicker response to demand changes.

(3) Reductions in setup costs.

(4) More efficient machine utilization.

(5) Quicker response to revisions to the master schedule.

Words/Phrases and Expressions

Computer Integrated Manufacturing (CIM) 计算机集成制造
interchangeable 互换的
just-in-time (JIT) 及时
product 产品
process 加工
design 设计
planning 计划,规划
shipping 运输
receive 接收
programme 程序
store 储存
finance 金融,财务
market 市场,营销
produce 生产,制造
manual manufacturing 手工制造
mechanization 机械化
specialization 专业化
intelligent 智能化
standardization 标准化
human resource 人力资源
strategic planning 战略规划
accounting 会计
distribution 分布

Chapter 11 Flexible Manufacturing Systems

11.1 Introduction

Today, the industrial world has become a true international marketplace. Our transportation networks have created a "world market" that we participate in on a daily basis. For any industrial country to compete in this market, it is necessary for a company to provide an economic, high-quality product to the market place in a timely manner. The importance of integrating product design and process design cannot be over emphasized. However, even once a design has been finalized, today's manufacturing industries must be willing to accommodate their customers by allowing for last-minute engineering design changes without affecting their shipping schedule or altering their product quality.

Today, the use of computers in manufacturing is as common as the use of computers in education. Manufacturing systems are being designed that not only process parts automatically, but also move the parts from machine to machine and sequence the ordering of operations in the system. Today, manufacturing systems used in industry employ a variety of automation features. Numerical control machines and flexible manufacturing systems (FMSs) have become a part of how durable goods are produced not only in the United States but throughout the world.

11.2 Flexible Manufacturing Systems

A flexible manufacturing system, or FMS as such a system is more commonly known, is a reprogrammable manufacturing system capable of producing a variety of products automatically. Since Henry Ford first introduced and modernized the transfer line, we have been able to perform a variety of manufacturing operations automatically. However, altering these systems to accommodate even minor changes in the product was at times impossible. Whole machines might have to be introduced to the system while other machines or components are retired to accommodate small changes in a product. In today's competitive market place, it is necessary to accommodate customer changes or the customer will find someone else who will accommodate his changes. Conventional manufacturing systems have been marked by one of two distinct features:

(1) Job-shop types of systems were capable of producing a variety of product but at a large cost.

(2) Transfer equipment could produce large volumes of a product at a reasonable cost, but

was limited to the production of one, two, or very few different parts.

The advent of numerical control and robotics has provided us with the basic processing capabilities to reprogram a machine's operation with minimum set-up time NC machines and robots provide the basic physical building blocks for reprogrammable manufacturing systems.

11.3　Equipment

11.3.1　Machines

In order to meet the requirements of the definition of a FMS, the basic processing in the system must be automated. Large, heavy parts require large, powerful handling systems such as roller conveyors, guided vehicles, or track-driven vehicle systems. The number of machines and the layout of the machines also present another design consideration. If a single material handler is to move parts to all the machines in the system, then the work envelope of the handler must be at least as large as the physical system. A robot is normally only capable of addressing one or two machines and a load and unload station. A conveyor or automatic guided vehicle (AGV) system can be expanded to include miles of factory floor. The material-handling system must also be capable of moving parts from one machine to another in a timely manner. The machines in the system will be unproductive if they spend much of their time waiting for parts to be delivered by the material handler. If many parts are included in the system and they require frequent visits to machines, then the material-handling system must be capable of supporting these activities. This can usually be accommodated either by using a very fast handling device or by using several devices in parallel.

11.3.2　Tooling and Fixtures

Versatility is the key to most FMSs, and as such, the tooling used in the system must be able of supporting a variety of products or parts. The use of special forming tools in a FMS is generally not employed in practice. The contours obtained by using forming tools can usually be obtained through a contour-control NC system and a standard mill. The standard mill can then be used for a variety of parts rather than to produce a single special contour. An economic analysis of the costs and benefit of any special tooling is necessary to determine the best tooling combination. However, since NC machines have a small number of tools that are accessible, very few special tools should be used.

One of the commonly neglected aspects of a FMS is the fixturing used in these systems. Since fixtures are part of the tooling of the system, one could argue that they should also be standard for the system. Work on creating "flexible fixtures" that could be used to support a variety of components has only begun recently. One unique aspect of many FMSs is that the part is also moved about the system in the fixture. Fixtures are made to the same dimension so that the material-handling system can be specialized to handle a single geometry. Parts are located precisely on the fixture and moved from one station to another on the fixture. Fixtures of this type are usually called pallet fixtures or pallets. Many of the pallet fixtures employed today have standard "T-slots" cut in them, and use standard fixture kits to create the part-locating and holding environment needed for machining.

11.3.3 Computer Control of Flexible Manufacturing Systems

An FMS is a complex network of equipment and activities that must be controlled via a computer or network of computers. In order to make the task of controlling an FMS more tractable, the system is usually decomposed in a task-based hierarchy. One of the standard hierarchies that have evolved is the National Bureau of Standards factory control hierarchy. This hierarchy consists of five levels and is illustrated in Figure 11.1. The system consists of the physical machining equipment at the lowest level of the system. Workstation equipment resides just above the process level and provides integration and interface functions for the equipment. For instance, pallet fixtures and programming elements are part of the workstation. The workstation typically provides both man-machine interface as well as machine-part interface. Off-line programming, such as APT for NC, or AML for a robot, would reside at the workstation level.

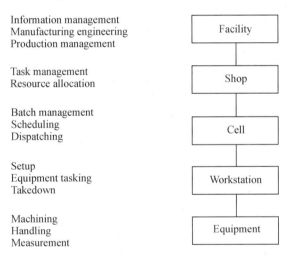

Figure 11.1 Control hierarchy used by the National Bureau of Standards. Elements of the data-driven control function within the NBS AMRF architecture: the facility, shop, cell, workstation, and equipment levels of control

The cell is the unit in the hierarchy where interaction between machines becomes part of the system. The cell controller provides the interface between the machines and the material-handling system. As such, the cell controller is responsible for sequencing and scheduling parts through the system. At the shop level, integration of multiple cells occurs, as well as the planning and management of inventory. The facility level is the place in the hierarchy where the master production schedule is constructed and manufacturing resource planning is conducted. Ordering of materials, planning inventories, and analyzing business plans are part of the activities that affect the production system. Poor business and manufacturing plans will incapacitate the manufacturing system just as surely as the unavailability of a machine.

Words/Phrases and Expressions

flexible manufacturing 柔性制造
Automatic Guided Vehicle (AGV) 自动引导车
fixture 夹具
workstation 工作站
information management 信息管理
facility 设施
task management 任务管理
resource allocation 资源配置
batch 批次
schedule 日程,时间安排
dispatch 派遣
hierarchy 层次

Chapter 12 Nanotechnology and Micro-Machine

12.1 Nanotechnology

12.1.1 Introduction

What is nanotechnology? It is a term that entered into the general vocabulary only in the late 70's, mainly to describe the metrology associated with the development of X-ray, optical and other very precise components. We defined nanotechnology as the technology where dimensions and tolerances in the range 0.1–100 nm (from the size of the atom to the wavelength of light) play a critical role.

This definition is too all-embracing to be of practical value because it could include, for example, topics as diverse as X-ray crystallography, atomic physics and indeed the whole of chemistry. So the field covered by nanotechnology is later narrowed down to manipulation and machining within the defined dimensional range (from 0.1 to 100 nm) by technological means, as opposed to those used by the craftsman, and thus excludes, for example, traditional forms of glass polishing. The technology relating to fine powders also comes under the general heading of nanotechnology, but we exclude observational techniques such as microscopy and various forms of surface analysis.

12.1.2 Materials Technology

The wide scope of nanotechnology is demonstrated in the materials field, where materials provide a means to an end are not an end in themselves. For example, in electronics, inhomogeneities in materials, on a very fine scale, set a limit to the nanometer-sized features that play an important part in semiconductor technology, and in a very different field, the finer the grain size of an adhesive, the thinner will be the adhesive layer, and the higher will be the bond strength.

1. Advantages of Ultra-Fine Powders

In general, the mechanical, thermal, electrical and magnetic properties of ceramics, sintered metals and composites are often enhanced by reducing the grain or fiber size in the starting materials.

Other properties such as strength, the ductile-brittle transition, transparency, dielectric coefficient and permeability can be enhanced either by the direct influence of an ultra-fine microstructure or by the advantages gained by mixing and bonding ultra-fine powders.

2. Applications of Ultra-Fine Powders

Important applications include:

(1) Thin films and coatings—the smaller the particle size, the thinner the coating can be.

(2) Chromatography—the increase in specific surface area associated with small particles, allows column lengths to be reduced.

(3) Electronic ceramics—reduction in grain size results in reduced dielectric thickness.

(4) Strength-bearing ceramics—strength increases with decreasing grain size.

(5) Cutting tools—smaller grain size results in a finer cutting edge, which can enhance the surface finish.

(6) Impact resistance—finer microstructure increases the toughness of high-temperature steels.

(7) Cements—finer grain size yields better homogeneity and density.

(8) Gas sensors—finer grain size gives increased sensitivity.

(9) Adhesives—finer grain size gives thinner adhesive layer and higher bond strength.

3. Precision Machining and Materials Processing

A considerable overlap is emerging in the manufacturing methods employed in very different areas such as mechanical engineering, optics and electronics. Precision machining encompasses not only the traditional techniques such as turning, grinding, lapping and polishing refined to the nanometer level of precision, but also the application of "particle" beams, ions, electrons and X-rays. Ion beams are capable of machining virtually any material and the most frequent applications of electrons and X-rays are found in the machining or modification of resist materials for lithographic purposes. The interaction of the beams with the resist material induces structural changes such as polymerization that alter the solubility of the irradiated areas.

(1) Diamond turning.

The large optics diamond-turning machine at the Lawrence Livermore National Laboratory represents a pinnacle of achievement in the field of ultra-precision machine tool engineering. This is a vertical-spindle machine with a face plate 1.6 m in diameter and a maximum tool height of 0.5 m. Despite these large dimensions, machining accuracy for form is 27.5 nm RMS and a surface roughness of 3 nm is achievable, but is dependent both on the specimen material and cutting tool.

(2) Grinding.

①Fixed abrasive grinding. The term "fixed abrasive" denotes that a grinding wheel is employed in which the abrasive particles, such as diamond, cubic boron nitride or silicon carbide, are attached to the wheel by embedding them in a resin or a metal. The forces generated in grinding are higher than in diamond turning and usually machine tools are tailored for one or the other process. Some Japanese work is in the vanguard of precision grinding, and surface finishes of 2 nm (peak-to-valley) have been obtained on single-crystal quartz samples using extremely stiff grinding machines.

②Loose abrasive grinding. The most familiar loose abrasive grinding processes are lapping and polishing where the workpiece, which is often a hard material such as glass, is rubbed against a softer material, the lap or polisher, with an abrasive slurry between the two surfaces. Loose abrasive grinding techniques can under appropriate conditions produce unrivalled accuracy both in form and surface finish when the workpiece is flat or spherical. Surface figures to a few nm and surface finishes to better than 0.5 nm may be achieved.

(3) Thin-film production.

The production of thin solid films, particularly for coating optical components, provides a good example of traditional nanotechnology. There is a long history of coating by chemical methods, electro-deposition, diode sputtering and vacuum evaporation, while triode and magnetron sputtering and ion-beam deposition are more recent in their wide application.

Because of their importance in the production of semiconductor devices, epitaxial growth techniques are worth a special mention. Epitaxy is the growth of a thin crystalline layer on a single-crystal substrate, where the atoms in the growing layer mimic the disposition of the atoms in the substrate.

12.1.3 Applications

There is an all-pervading trend to higher precision and miniaturization, and to illustrate this a few applications will be briefly referred to in the fields of mechanical engineering, optics and electronics.

1. Mechanical Engineering

One of the earliest applications of diamond turning was the machining of aluminum substrates for computer memory discs. Some precision components falling in the manufacturing tolerance band of 5-50 nm include gauge blocks, diamond indenter tips, microtome blades.

2. Optics

The large demand for infrared optics from the 1970s onwards could not be met by the traditional optics manufacture suppliers, and provided a stimulus for the development and application of diamond-turning machines to optic manufacture. The technology has now progressed and the surface figure and finishes that can be obtained span a substantial proportion of the nanotechnology range. Important applications of diamond-turned optics are in the manufacture of unconventionally shaped optics, for example, axicons and more generally, aspherics and particularly off-axis components, such as paraboloids.

3. Electronics

In semiconductors, nanotechnology has long been a feature in the development of layers parallel to the substrate and in the substrate surface itself, and the need for precision is steadily increasing with the advent of layered semiconductor structures. About one quarter of the entire semiconductor physics community is now engaged in studying aspects of these structures. Normal to the layer surface, the structure is produced by lithography, and for research purposes at least, nanometer-sized features are now being developed using X-ray and electron-and ion-beam techniques.

12.2 Micro-Machine

12.2.1 Introduction

From the beginning, mankind seems instinctively to have desired large machines and small machines. That is, "large" and "small" in comparison with human-scale. Machines larger than human are powerful allies in the battle against the fury of nature; smaller machines are loyal partners that do whatever they are told.

If we compare the facility and technology of manufacturing larger machines, common sense tells us that the smaller machines are easier to make. Nevertheless, throughout the history of technology, larger machines have always stood out. The size of the restored models of the water-mill invented by Vitruvius in the Roman Era, the windmill of the Middle Ages, and the steam engine invented by Watt is overwhelming. On the other hand, smaller machines in history of technology are mostly tools. If smaller machines are easier to make, a variety of such machines should exist, but until modern times, no significant small machines existed except for guns and clocks.

This fact may imply that smaller machines were actually more difficult to make.

Of course, this does not mean simply that it was difficult to make a small machine. It means that it was difficult to invent a small machine that would be significant to human beings.

Some people might say that mankind may not have wanted smaller machines. This theory, however, does not explain the recent popularity of palm-size mechatronics products.

The absence of small machines in history may be due to the extreme difficulty in manufacturing small precision parts.

12.2.2 Micro-Electronics and Mechatronics

The concept of micromachines and related technologies is still not adequately unified, as these are still at the development stage. The micromachines and related technologies are currently referred to by a variety of different terms. In the United States, the accepted term is "Micro Electro Mechanical Systems" (MEMS). In Europe, the term "Microsystems Technology" (MST) is common, while the term "micro-engineering" is sometimes used in Britain, meanwhile in Australia "micro-machine". The most common term if it is translated into English is "micromachine" in Japan. However, "micro-robot" and "micro-mechanism" are also available case by case.

The appearance of these various terms should be taken as reflecting not merely diversity of expression, but diversity of the items referred to. Depending on whether the item referred to is an object or a technology, the terminology may be summed up as follows.

Object: micro-robot, micro-mechanism.

Technology: micro-engineering, MST.

Object & technology: MEMS, micro-machine.

MEMS and MST stem from mechatronics, and have developed dealing mainly with machine systems. In this sense, MEMS and MST on the one hand and micromachines and micro-engineering, on the other hand form two separate groups, but as the former has started to move in the direction of machine systems, while the latter has already incorporated microelectronics, the differences between the two groups are gradually disappearing.

12.2.3 The Evolution of Machines and Micromachines

Many researchers see micromachines as the ultimate in mechatronics, developed out of machine systems.

Ever since the Industrial Revolution, machine systems have grown larger and larger in the course of their evolution. Only very recently has evolution in the opposite direction begun, with the appearance of mechatronics. Devices such as video cameras, tape recorders, portable telephones, portable copiers which at one time were too large to put one's arms around, now fit on the palm of one's hand.

Miniaturization through mechatronics has resulted mainly from the development of electronic controls and control software for machine systems, but the changes to the structural parts of machine systems have been minor compared to those in the control systems. The next target in miniaturization of machine systems is miniaturization of the structural parts left untouched by present mechatronics. These are the micromachines which are seen as the ultimate in mechatronics. Seen in this light, the aim of micromachines can be expressed as follows.

"Micromachines are autonomous machines which can be put on a fingertip, composed of parts the smallest sized of which is a few dozen micrometers".

That is, since micromachines which can be put on a fingertip have to perform operations in spaces inaccessible to humans, they are required to be autonomous and capable of assessing situations independently, as are intelligent robots. To achieve this kind of functionality, a large number of parts must be assembled in a confined space. This factor determines the size of the smallest parts, and given the resolution of micro-machining systems, a target size of several dozen micrometers should be achievable.

Words/Phrases and Expressions

nanotechnology 纳米技术
micro-machine 微机械
metrology 计量学
X-ray X射线

dimension 尺寸
tolerance 公差
all-embracing 包罗万象，无所不能
crystallography 晶体学
atomic physics 原子物理
inhomogeneity 多相(性)
nanometer-sized 纳米级
ultra-fine powder 超细粉末
ductile-brittle transition 韧性-脆性转变
transparency 透明度
dielectric coefficient 介电常数
permeability 渗透性
thin film 薄膜
coat 涂层
chromatography 色谱法
lithographic 光刻
polymerization 聚合
irradiate 照射
solubility 溶解度
boron nitride 氮化硼
silicon carbide 碳化硅
electro-deposition 电沉积
diode 二极管
sputter 喷射
vacuum 真空
evaporation 蒸发
triode 三极管
magnetron 磁控
ion-beam 离子束
epitaxial 外延的
crystalline 结晶的
single-crystal 单晶
substrate 基体
mimic 模仿
disposition 配置
all-pervading 普遍的
optics 光学
axicon 轴棱锥

aspheric 非球面的,非球面镜头
paraboloid 抛物面
electron-beam 电子束
Micro Electro Mechanical Systems (MEMS) 微机电系统
Microsystems Technology (MST) 微系统技术

Chapter 13　Biofabrication

13.1　Definition and Scope of Biofabrication

Biofabrication can be defined as the production of complex living and non-living biological products from raw materials such as living cells, molecules, extracellular matrices, and biomaterials. Cell and developmental biology, biomaterials science, and mechanical engineering are the main disciplines contributing to the emergence of biofabrication technology.

Biofabrication is a technology, as opposed to a basic science, and is part of the much broader field of biotechnology. The prefix "bio" implies that either raw materials, or processes, or final products (or all these) must be biology inspired or biology based. The term "fabrication" means making or constructing something from a raw or semi-finished material or creating something that is different from its components. In this context, biofabrication deals with science, engineering and technology or production, based on using living matter as raw materials.

Biofabrication can be narrowly defined as the production of complex biological products using living cells, molecules, extracellular matrices, and engineered biomaterials. This definition does limit biofabricated products to living tissues or organs, it implies that raw materials must include living or bio-inspired matter, component processes must be biology based and intermediate products must be complex living tissue constructs. However, according to a more inclusive definition, biofabrication encompasses a broad range of physical, chemical, biological, and/or engineering processes with various applications in tissue science and engineering, disease pathogenesis, and drug pharmacokinetic studies, biochips and biosensors, cell printing, patterning and assembly, and emerging organ printing.

Biology, mechanical engineering (CAD/CAM, additive manufacturing), and materials science (primarily biomaterials) constitute the main disciplines and basic technological pillars of biofabrication (Figure 13.1).

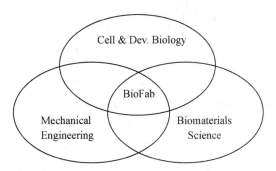

Figure 13.1 The main disciplines contributing to the emergence of biofabrication: cell and developmental biology, mechanical engineering, and biomaterials science

13.2 Practical Applications of Biofabrication Technologies

1. Biofuel Production from Algae

Energy consumption has a strong linear correlation with the economic growth of a nation, which makes sustainable energy production a top priority for any country. Algae growing at first in a perfused closed bioreactor and then in open ponds can be a most effective method for biofuel production. According to some estimates it will be 300 times more effective than a leading plant source for biofuel production such as palm oil.

2. Biofabrication of Human Tissues and Organs for Implantation

One of the most obvious and highly desirable practical applications of biofabrication technology is bioengineering of living human tissues and organs suitable for implantation. Advanced biofabrication technology can help to design cost-effective industrial production of living human organs or living and implantable organ constructs. Investment in artificial human organ biofabrication technologies is both economically and socially justifiable and morally sound.

The exciting new strategy, however, is likely to revolutionize the treatment of patients who need new vital structures: the creation of man-made tissues or organs, known as neo-organs. The groundbreaking applications involve fabricated skin, cartilage, bone, ligament and tendon and make musings of "off-the-shelf" whole organs, which seems somewhat fantastic.

Promoting tissue and organ development via growth factors is obviously a considerable step forward. This approach is now actually in use in some patients, notably those with skin wounds or cartilage damage. Charles A. Vacanti of the University of Massachusetts Medical School in Worcester, has shown that new cartilage can be grown in the shapes of ears (Figure 13.2), noses and other recognizable forms. A group of scientists headed by Michael V. Sefton at the University of Toronto began an ambitious project to grow new hearts for the multitude of people who would die from heart failure every year.

Figure 13.2　A tissue-engineered human-ear mouse

3. Biosensors and Bioreporters in Space Research

Although the USA, China and Russia are still actively pursuing rather expensive and risky human space exploration and manned space trips, the European Union is more focused on an alternative approach-unmanned automatic space exploration. However, even in the case of planned Moon and Mars manned missions, the development of sophisticated cell and tissue-based bioreporters and biosensors is essential to check radiation safety. Thus, future applications of biofabrication in space research will probably be related to designing, fabricating, and testing sophisticated miniaturized tissue-based radiation biosensors and bioreporters. One such sophisticated microfluidic bioreporter has already been developed for NASA. The most challenging goal is to create an in vitro tissue analog of human organisms or organs which mimic the essential aspects of complexity of human organisms, including the radiation sensitive immune system with circulating lymphocytes and stem cells.

Biofabrication represents a potentially powerful technological platform to support sustainable manufacturing by creating new and transforming existing industries. Biofabrication could be the dominant paradigm for 21st century manufacturing.

Words/Phrases and Expressions

biofabrication 生物制造
extracellular matrices 细胞外基质
biomaterial 生物材料
cell 细胞
developmental biology 发育生物学
engineered 工程化
tissue construct 活体组织结构体
pathogenesis 发病机理
pharmacokinetic 药物代谢动力学
perfuse 灌注
bioreactor 生物反应器
biofuel 生物燃料
palm oil 棕榈油
neo-organ 新型器官
cartilage 软骨
ligament 韧带

tendon 肌腱
muse 冥思苦想
off-the-shelf 现成的
biosensors 生物传感器
microfluidic 微流体
bioreporter 生物感应器
vitro tissue analog 体外组织模拟

Chapter 14 Advanced Manufacturing Mode

The advanced manufacturing mode is an effective production method and a certain production organization form in the manufacturing industry in order to improve product quality, market competitiveness, production scale and production speed, and finally complete specific production tasks. Typical advanced manufacturing modes include: agile manufacturing (AM), lean production (LP), concurrent engineering (CE), virtual manufacturing (VM), rapid response manufacturing system (RMS), green manufacturing (GM), etc..

14.1 Agile Manufacturing

Manufacturing industry may well be on the verge of a major paradigm shift. This shift is likely to take us away from mass production, way beyond lean manufacturing, into a world of Agile Manufacturing.

14.1.1 Definition and Concepts

Agile Manufacturing should primarily be seen as a business concept. Its aim is quite simple—to put our enterprises way out in front of our primary competitors.

In Agile Manufacturing our aim is to develop agile properties. We will then use this agility for competitive advantage, by being able to rapidly respond to changes occurring in the market environment and through our ability to use and exploit a fundamental resource-knowledge. Agile Manufacturing enterprises are expected to be capable of rapidly responding to changes in customer demand.

One fundamental idea in the exploitation of this resource is the idea of using technologies to lever the skills and knowledge of our people. We need to bring our people together, in dynamic teams formed around clearly identified market opportunities, so that it becomes possible to lever one another's knowledge. Through these processes we should seek to achieve the transformation of knowledge and ideas into new products and services, as well as improvements to our existing products and services.

The concept of Agile Manufacturing is also built around the synthesis of a number of enterprises that each have some core skills or competencies which they bring to a joint venturing operation, which is based on using each partner's facilities and resources.

14.1.2 Some Key Issues in Agile Manufacturing

1. The "I am a Horse" Syndrome

There is an old saying that hanging a sign on a cow that says "I am a horse" does not make it a horse. There is a real danger that Agile Manufacturing will fall prey to the unfortunate tendency in manufacturing circles to follow fashion and to relabel everything with a new fashionable label.

The dangers in this are two folds. First, it will give Agile Manufacturing a bad reputation. Second, instead of getting to grips with the profound implications and issues raised by Agile Manufacturing, management will only acquire a superficial understanding, which leaves them vulnerable to those competitors that take Agile Manufacturing seriously. Of course this is good news for the competitors.

One sure way to fail with Agile Manufacturing is to hang a new sign up. Get smart, resist the temptation, and put the paint brush away.

2. The Existing Culture of Manufacturing

One of the important things that is likely to hold us back from making a quantum leap forward and exploring this new frontier of Agile Manufacturing, is the baggage of our traditions, conventions and our accepted values and beliefs. A key success factor is, without any doubt, the ability to master both the soft and hard issues in change management. However, if we are to achieve agility in our manufacturing enterprises, we should first try to fully understand the nature of our existing cultures, values, and traditions.

We need to achieve this understanding, because we need to begin to recognize and come to terms with the fact that much of what we have taken for granted, probably no longer applies in the world of Agile Manufacturing. Achieving this understanding is the first step in facing up to the pain of consigning our existing culture to the garbage can of historically redundant ideas.

3. Understanding Agility

Agility is defined in dictionaries as quick moving, nimble and active. This is clearly not the same as flexibility which implies adaptability and versatility. Agility and flexibility are therefore different things.

Leanness (as in lean manufacturing) is also a different concept to agility. Sometimes the terms lean and agile are used interchangeably, but this is not appropriate.

The term lean is used because lean manufacturing is concerned with doing everything with less. In other words, the excess of wasteful activities, unnecessary inventory, long lead times, etc are cut away through the application of just-in-time manufacturing, concurrent engineering, overhead cost reduction, improved supplier and customer relationships, total quality management, etc..

We can also consider CIM in the same light. When we link computers across applications, across functions and across enterprises we do not achieve agility. We might achieve a necessary condition for agility, that is, rapid communications and the exchange and reuse use of data, but we do not achieve agility.

Thus agility is not the same as flexibility, leanness or CIM. Understanding this point is very important. But if agility is none of these things, then what is it? This is a good question, and not one easily answered. Yet most of us would recognize agility if we saw it.

14.2 Lean Manufacturing

14.2.1 What Is Lean Manufacturing?

Lean Manufacturing is a unified, comprehensive set of philosophies, rules, guidelines, tools, and techniques for improving and optimizing discrete processes.

While Lean was born in large volume, repetitive manufacturing for the automotive industry sector, Lean principles and benefits apply to all processes (health care, service, high tech, salesand marketing, fast food, etc.). For this reason, some call it "Lean Thinking", rather than the more restrictive title of "Lean Manufacturing".

14.2.2 History of Lean Manufacturing

Lean Manufacturing started as the Toyota Production System (TPS), developed by the Toyoda (now Toyota) Motor Car Company. The tools and techniques started to emerge that allowed Toyota to achieve its TPS goals. Toyota's engineers looked to Henry Ford (inventor of the assembly line), Taylor (inventor of Modern Management techniques and Industrial Engineering), and Dr. W. Edwards Deming (Father of Modern Quality Management). Based on these early beginnings, the techniques were refined, honed, and improved in all areas.

With the invasion of the North American market by Volkswagen in the 1960s, and Toyota in the 1970s, and a world-wide recession, the American automotive industry was in for major changes and de-stabilization.

A 5 year, $5 million research project by MIT was started so as to analyze the world-wide automotive industry. Dr. Womack coined the phrase "Lean Manufacturing" and explained how Toyota could manufacture a car, ship it to North America, and sell it faster and cheaper than domestically made vehicles. The Japanese vehicles innovated at an extremely rapid rate, while N. American designed and built vehicles tended to change at a very slow rate.

Today, Lean Thinking (re-coined again so as to signal that the same techniques can be used in banks, service organizations, hospitals, and all manner of business systems) is being used world-wide in a growing number of organizations. It is applied at the point of contact with customers, as well as back room work. It applies to Engineering and Design offices, as well as traffic flow in urban centres.

14.2.3 Some Key Issues in Agile Manufacturing

1. Benefits of Lean Manufacturing

(1) Overhead operating costs reduced.

(2) Sales higher.

(3) Profits increased.

(4) Lead time cut.

(5) Process queues cut.

(6) Less frustration on-the-job.

2. Principles of Lean Manufacturing

(1) Voice of the customer.

(2) Continuous improvement.

(3) Recognizeand eliminate waste everywhere.

(4) Over-production (more, earlier, and faster).

(5) Inventory (more than 1-piece flow).

(6) Defects (non-zero defect rates).

(7) Over-processing (low or no value added by unnecessary work).

(8) Waiting.

(9) People's talents and motivations.

(10) Motion.

(11) Transportation.

3. Toolsand Techniques of Lean Manufacturing

(1) Value mapping.

(2) SMED (Single Minute Exchange of Dies).

(3) Piece flow.

(4) KanBan (Inventory Control via Card System).

(5) Poka yoke (Mistake Proofing) and jidoka.

(6) 5S (Separate, Simplify, Standardize, Sustain, Straighten).

(7) Total productive maintenance.

(8) Visual management.

(9) Line optimization.

(10) Synchronous manufacturing.

4. How is Lean Manufacturing Achieved?

(1) Assessing the current process.

(2) Understanding the customer's true desires and future market trends.

(3) Trainingand buy-in by Sr. management.

(4) Developing profound knowledge of the manufacturing process.

(5) Applying lean tools and techniques at the most critical processes.

(6) Spreading out the lean implementation to all auxiliary areas until a fully integrated manufacturing process is obtained.

(7) Implement lean with suppliers.

(8) Implement lean with downstream supply chain organizations, including customers.

(9) Apply lean into off-line and non-manufacturing areas (Engineering, Design, Marketing, etc.).

14.3 Concurrent Engineering

14.3.1 What Is Concurrent Engineering?

Concurrent engineering is a business strategy which replaces the traditional product development process with one in which tasks are done in parallel and there is an early consideration for every aspect of a product's development process. This strategy focuses on the optimization and distribution of a firm's resources in the design and development process to ensure effective and efficient product development process.

14.3.2 How to Apply Concurrent Engineering?

1. Commitment, Planning and Leadership

Concurrent engineering is not a trivial process to apply. If firms are going to commit to concurrent engineering then they must first devise a plan. This plan must create organizational change throughout the entire company or firm.

There must be a strong commitment from the firm's leadership in order to mandate the required organizational changes from the top down. Concurrent engineering without leadership will have no clear direction or goal. On the other hand, concurrent engineering with leadership, management support, and proper planning will bring success in today's challenging market place.

2. Continuous Improvement Process

On current engineering is not a one size fits all solution to a firm's development processes. There are many different aspects of concurrent engineering which may or may not fit in a corporation's development process. Concurrent engineering is only a set of process objectives and goals that have a variety of implementation strategies. Therefore, concurrent engineering is an evolving process that requires continuous improvement and refinement.

This continuous improvement cycle consist of planning, implementing, reviewing and revising. The process must be updated and revised on a regular basis to optimize the effectiveness and benefits in the concurrent engineering development process.

3. Communication and Collaboration

The implementation of concurrent engineering begins by creating a corporate environment that facilitates communication and collaboration not just between individuals, but also between separate organizations and departments within the firm. This may entail major structural changes, re-education of the existing work-force, and/or restructuring of the development process.

14.3.3 Basic Principles of Concurrent Engineering

(1) Get a strong commitment from senior management.

(2) Establish unified project goals and a clear business mission.

(3) Develop a detailed plan early in the process.

(4) Continually review your progress and revise your plan.

(5) Develop project leaders that have an overall vision of the project and goals.

(6) Analyze your market and know your customers.

(7) Suppress individualism and foster a team concept.

(8) Establish and cultivate cross-functional integration and collaboration.

(9) Transfer technology between individuals and departments.

(10) Break project into its natural phases.

(11) Develop metrics.

(12) Set milestones throughout the development process.

(13) Collectively work on all parts of project.

(14) Reduce costs and time to market.

(15) Complete tasks in parallel.

14.3.4 When do Companies Use Concurrent Engineering?

The majority of a product's costs are committed very early in the design and development process. Therefore, companies must apply concurrent engineering at the onset of a project. This makes concurrent engineering a powerful development tool that can be implemented early in the conceptual design phase where the majority of the products costs are committed.

There are several applications in which concurrent engineering may be used. Some primary applications include product research, design, development, re-engineering, manufacturing, and redesigning of existing and new products. In these applications, concurrent engineering is applied throughout the design and development process to enable the firm to reap the full benefits of this process.

14.3.5 Why do Companies Use Concurrent Engineering?

1. Competitive Advantage

The reasons that companies choose to use concurrent engineering is for the clear cut benefits and competitive advantage that concurrent engineering can give them. Concurrent engineering can benefit companies of any size, large or small. While there are several obstacles to initially implementing concurrent engineering, these obstacles are minimal when compared to the long term benefits that concurrent engineering offers.

2. Increased Performance

Companies recognize that concurrent engineering is a key factor in improving the quality, development cycle, production cost, and delivery time of their products. It enables the early discovery of design problems, thereby enabling them to be addressed up front rather than later in the development process. Concurrent engineering can eliminate multiple design revisions, prototypes, and re-engineering efforts and create an environment for designing right the first time.

3. Reduced Design and Development Times

Companies that use concurrent engineering are able to transfer technology to their markets and customers more effectively, rapidly and predictably. They will be able to respond to customers' needs and desires, to produce quality products that meet or exceeds the consumer's expectations. They will also be able to introduce more products and bring quicker upgrades to their existing products through concurrent engineering practices. Therefore companies use concurrent engineering to produce better quality products, developed in less time, at lower cost, that meets the customer's needs.

Words/Phrases and Expressions

Advanced Manufacturing Mode 先进制造模式
Agile Manufacturing (AM) 敏捷制造
Lean Production (LP) 精益生产
Concurrent Engineering (CE) 并行工程
Virtual Manufacturing (VM) 虚拟制造
Rapid Response Manufacturing System (RMS) 快速响应制造系统
Green Manufacturing (GM) 绿色制造
syndrome 综合症
superficial 肤浅的
key issue 关键问题
over-production 过度生产
inventory 库存
defect 缺陷
line optimization 线路优化
synchronous 同步的
plan 计划
implement 实施
review 审查
revise 修订
communication 沟通
collaboration 合作
restructure 重组
milestone 里程碑

Chapter 15   Environmentally Conscious Design and Manufacturing (ECD&M)

15.1   Introduction

Industrial countries are beginning to face one of the consequences of the rapid development of the last decade. Wide diffusion of consumer goods and shortening of product lifecycles have caused an increasing quantity of used products being discarded.

Facing environmental problem, both the government and industrial companies are making more strict regulations to promote environmentally friendly products and technology. All of the regulations intend to minimize the environmental impact of products. Products affect the environment at many points in their lifecycles. These environmental effects result from the interrelated decisions made at various stages of a product's life. Once a product moves from the drawing board into the production line, its environmental attributes are largely fixed. Therefore, it is necessary to support the design function with tools and methodologies that enable an assessment of the environmental consequences (such as emissions, exposure, and effects) in each phase.

Environmentally conscious design and manufacturing (ECD&M) is a view of manufacturing that includes the social and technological aspects of the design, synthesis, processing, and use of products in continuous or discrete manufacturing industries.

The benefits of ECD&M include safer and cleaner factories, worker protection, reduced future costs for disposal, reduced environmental and health risks, improved product quality at lower cost, better public image, and higher productivity. Environmentally conscious technologies and design practices will also allow manufacturers to minimize waste and to turn waste into a profitable product.

15.2   Overview

Although manufacturing industries contribute significantly to prosperity, they also generate approximately 5.5 billion tons of non-hazardous waste and 0.7 billion tons of hazardous waste each year. Historically, much effort focused on the proper treatment and disposal of toxic and hazardous waste from industries. Unfortunately, this reactive environmental protection approach cannot completely solve the problems of potential toxic or hazardous materials releasing from products or the waste stream into the environment. To effectively protect the environment, pollution control must be incorporated into every aspect of manufacturing.

As opposed to the traditional "end-of-pipe" treatment for pollution control, ECD&M is a proactive approach to minimize the product's environmental impact during its design and manufacturing, and thus to increase the product's competitiveness in the environmentally conscious market place.

There are two general approaches to ECD&M. In the first approach (zero-wasted lifecycle), it is assumed that the environmental impact of a product during its lifecycle can be reduced to zero. The cycle can be absolutely sustainable, and the product may be designed, manufactured, used, and disposed of without affecting the environment. The emphasis in this approach is to create a product cycle that is as sustainable as possible. Sustainable production means that products are designed, produced, distributed, used and disposed of with minimal (or none) environmental and occupational health damages, and with minimal use of resources (material and energy). The sustainability of a system can be considered as the ability of that system to be maintained or prolonged.

The second approach (incremental waste lifecycle control) is based on the premise that there is a certain amount of negative impact from the current process cycle. This impact can be reduced or cleaned based on some improvement in technology that is named as incremental waste lifecycle control. This approach is to reduce the negative impact of hazardous materials through clean technology. A "cleaner technology" is a source reduction or recycling method applied to eliminate or significantly reduce hazardous waste generation. Research on ECD&M can be categorized into two areas, namely, environmentally conscious product design and environmentally conscious process design, also called environmentally conscious manufacturing (ECM). A hierarchy comprised of environmentally conscious products is shown in Figure 15.1.

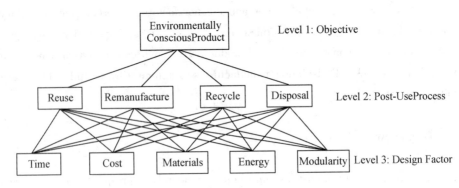

Figure 15.1 Hierarchy for Designing an Environmentally Conscious Product

At the first level of the hierarchy, the overall objectives for the system are considered when creating an environmentally conscious product.

At the second level, the four groups represent a post-use process that can be employed to achieve the objectives.

The third level consists of the five design factors that can facilitate the post-use processes and in turn accomplish the overall goal. This hierarchy shows the method of retiring products, whether the designers intend to have the product discarded in a landfall, or whether they plan to reuse or recycle part or the entire product.

The principle of ECM is to adopt those processes that reduce the harmful environmental impacts of manufacturing, including minimization of hazardous waste and emissions reduction of energy consumption, improvement of materials utilization efficiency, and enhancement of operational safety. Sandia National Laboratories' Environmentally Conscious Manufacturing Programs Department describes ECM as "the deliberate attempt to reduce ecological impacts of industrial activity without sacrificing quality, cost, reliability, performance, or energy utilization efficiency". The activities of ECM emphasize largely extracting the useful product from raw materials, the avoiding of waste generation at the source, or using waste to create other products. In addition, ECM involves refining operating procedures, replacing existing processes and developing new, waste-free processes, finding innovative ways to redesign products, and increasing recycling.

15.3 Environmental Engineering

All aspects of environmental problems are considered in environmental engineering, such as water and wastewater, environmental hydrology, environmental hydraulics and pneumatics, air, solid waste, noise, environmental modeling, and hazardous waste. Sincero defined environmental engineering as "the application of engineering principles, under constraint, to the protection and enhancement of the quality of the environment and to the enhancement and protection of public health and welfare". As the US environmental policy expanded from clean air to cradle-to-grave solid and hazardous waste management, environmental engineering research helped us better understand how pollutants migrate through soils, groundwater, and air, and developed treatment technologies to minimize their impact on natural and human environment.

The water resource management system includes water pollution, wastewater disposal, and the measurement of water quality, supply, and treatment. Crook presented guidelines for water reuse. These guidelines were developed to encourage and facilitate the orderly planning, design, and implementation of water reclamation. The air resource management system includes air pollution control and the measurement of air quality. The solid waste management system includes solid waste collection and landfill design. Williams indicated that source reduction, recycling and composting, waste-to-energy facilities, and landfills are the four basic approaches to waste management.

15.3.1 Pollution Prevention

The term "pollution prevention" is based on the technological and management advances program. The purposes of this program are: to reduce environmental releases; to lower costs in production from previous methods associated with pollution. The Pollution Prevention Act defined pollution prevention as "source reduction". Considering this definition, it may infer that the creation of pollutants may be reduced or eliminated through increased efficiency in the use of raw materials, energy, water, or other resources, or protection of natural resources by conservation. Pollution prevention is described as a "waste management hierarchy".

There are four preferences in the waste management hierarchy. The highest preference of the hierarchy is to reduce waste at the source of generation through the use of less toxic raw material, equipment changes, process redesign, better house-keeping, and materials management. The second preference is reuse and recycling of wastes that cannot be reduced at the source. The third preference is waste treatment, and the least preferred alternative is disposal. Two methods of source reduction can be used, product changes and process changes. These two methods reduce the volume and toxicity of production wastes and end products during their life-cycles.

The pollution prevention techniques used in industry are waste minimization and clean technology. Waste minimization includes source reduction and environmentally sound recycling. Source reduction is defined as many practices that reduces the amount of any hazardous substance, pollutant, or contaminant entering any waste stream or otherwise released into the environment prior to recycling, treatment, or disposal. Figure 15.2 shows source reduction methods. Clean technology uses less raw materials, energy, and water, generates less or no waste (gas, liquid, and solid), and recycles waste as useful materials in a closed system. The clean technology used in pollution prevention can be categorized into five groups: improved plant operations, in-process recycling, process modification, materials and product substitutions, and material separations.

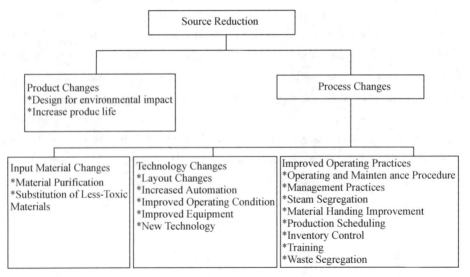

Figure 15.2 Source reduction methods

15.3.2 Design for Environment

Design for Environment (DFE) is defined by Lenox, Jordan, and Ehrenfeld as "the systematic process by which firms design products and processes in an environmentally conscious way". Another definition provided by Fiksel and Wapman is "the systematic consideration during new production and process development of design issues associated with environmental safety and health over the full product life cycle". The scope of DFE encompasses many disciplines, including environmental risk management, product safety, occupational health and safety, pollution prevention, ecology, resource conservation, accident prevention, and waste management. Horvath et al. provided three main goals of DFE: minimize the use of nonrenewable resources; effectively manage renewable resources; minimize toxic release to the environment. The elements of DFE include: metrics, practices, and analysis methods.

DFE requires the coordination of several design and data-based activities such as environmental impact metrics, data and database management, and design optimization (including cost assessments). The environmental metric is defined as "an algorithmic interpretation of levels of performance within an environmental criterion". The environmental criterion is the environmental attribute of the product, that is, the energy to heat water for a specific function, grams of CO_2 produced to deliver the above energy, chemical oxygen demand generated in the wastewater, degree of risk of exposure to a toxic substance, and so on. The product stewardship metrics include material conservation and waste reduction, energy efficiency, and design for environmental and manufacturing process emissions.

15.3.3 Lifecycle Engineering and Lifecycle Assessment

Lifecycle engineering (LCE) may also be referred to as lifecycle design (LCD). An outstanding analysis of lifecycle design that provides design support from the environmental point of view. Lifecycle design is based on the early product concept, including product and market research, design phases, manufacturing process, qualification, reliability issues, customer service, maintainability, and supportability issues. There are six phases in the product lifecycle: need recognition, design development, production, distribution, use, and disposal. All of the phases must be considered during the conceptual stage, where it is possible to inexpensively change solutions to accommodate the requirements in each phase and in the total lifecycle.

Lifecycle assessment is a family of methods for assessing materials, services, products processes, and technologies over the entire product life. The definition of product lifecycle assessment, developed by the Society of Environmental Toxicology and Chemistry, is as follows.

Lifecycle assessment is an objective process to evaluate the environmental burdens associated with a product or activity by identifying and quantifying energy and materials used and wastes released to the environment, to access the impact of those energy and material uses and releases to the environment, and to evaluate and implement opportunities to affect environmental improvements. The assessment includes the entire lifecycle of the product, process, or activity, encompassing extracting and processing raw materials, manufacturing, transportation and distribution, use, reuse, maintenance, recycling and final disposal.

Four phases of the product lifecycle are: product definition; product development; product manufacturing and marketing; product usage. At each of these phases there exists a definition of objectives, activities, and deliverables for the next phase.

15.3.4　Green Product Design

Green product design is expanded from pollution prevention. Green products—products that can reduce the burden on the environment during use and disposal have additional marketing appeal. Green product design refers to green engineering design, defined as "the study of and an approach to product and process evaluation and design for environmental compatibility that does not compromise products' quality or function". This approach is comprised of two parts: the evaluation of designs to assess their environmental compatibility and the relationship between design decisions and the green indicators. The aim of green engineering design is to develop an understanding of how design decisions affect a product's environmental compatibility.

Words/Phrases and Expressions

Environmentally Conscious Designand Manufacturing (ECD&M) 具有环境意识的设计和制造
emission 排放
exposure 暴露
hydrology 水文学
ecology 生态
resource 资源
conservation 节约,保护
nonrenewable 不可再生的
renewable 可再生的
metric 度量标准
practice 实践
analysis method 分析方法
coordination 协调
algorithmic 算法
interpretation 解释
lifecycle engineering (LCE) 生命周期工程
lifecycle design (LCD) 生命周期设计
reliability 可靠性
need recognition 需求意识
toxicology 毒理学
compatibility 兼容性

Chapter 16　Manufacturing Technology

Materials are processed to change their forms so that their economic value is increased. The company can then sell the materials or finished products for a profit.

Primary manufacturing companies change the forms of raw materials into industrial materials with standard sizes, shapes, and weights standard stock. Secondary manufacturing companies change the standard forms of industrial materials into one-piece products or components of multiple-piece products using materials forming and materials separating processes. Finally, secondary manufacturing companies use materials combining processes to turn components of multiple. Piece products into subassemblies and finished final assemblies.

16.1　Form-Changing Processes

Materials processing occurs in the production department of a manufacturing company. Production workers use tools and equipment to change the forms of metallic, ceramic, polymeric, and composite materials to increase their economic value. Because of this transition, the new forms of materials (or finished products) can be sold by the manufacturing company for a much higher amount of money. In this way, the company can make a profit. This is the primary goal of any manufacturing company.

Materials processing occurs in three distinct stages: the raw material stage, the standard stock stage, and the finished product stage.

All raw materials that are used in the manufacturing system are obtained in a natural state from the air, earth, or water. Plant raw materials (polymers) are typically acquired through a process called harvesting. Animal raw materials (polymers) are secured through the process of catching wild animals or slaughtering domesticated animals. Mineral raw materials (metals, mineral fuels, or ceramics) are acquired through mining and drilling processes.

Once raw materials are obtained, they must be refined to separate wanted materials from unwanted materials. After raw materials are refined, they are converted into industrial materials that have standard shapes, sizes, and weights. These standard-sized and-shaped industrial materials are also called standard stock. Examples of standard stock include bolts of fabric; ingots of pig iron, steel, gold, or lead; sheets or rolls of paper and steel; coils of wire; sheets of plywood, hardboard, or particleboard; granules or pellets of plastic resin; and bags of powdered flour or cement.

Many different kinds of processes can change refined raw materials into standard stock. For example, primary metals such as pig iron, steel, aluminum, or copper are converted into ingots, sheets, and coils using melting, casting, and rolling processes. Standard stock items almost always have to be further processed so that they can become useful consumer products.

16.2　Primary and Secondary Manufacturing

Manufacturing companies can be classified into primary manufacturing companies and secondary manufacturing companies. Primary manufacturing companies obtain and refine raw materials and produce standard forms of industrial materials. Examples of primary manufacturing companies include those producing metals (such as steel, aluminum, copper), textiles, lumber, glass, paper, and petroleum-based products or chemicals. The standard stock produced by these companies can be sold directly to individual consumers, or it can become material inputs to construction or secondary manufacturing companies, Figure 16.1.

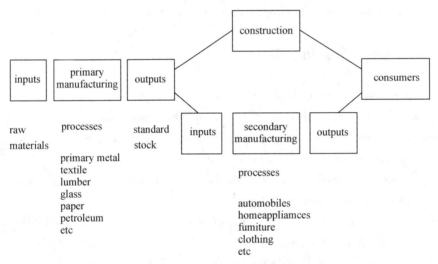

Figure 16.1　Primary manufacturing converts raw materials into standard stock, while secondary manufacturing converts standard stock into finished consumer products

Secondary manufacturing companies convert standard stock materials into products that can be bought by consumers in local stores. Product outputs of secondary manufacturing companies include such items as televisions, kitchen appliances, telephones, automobiles, medicines, furniture, and clothing. Secondary manufacturing companies first convert the standard stock purchased from primary industries into single-piece products or components of multiple-piece products using materials-forming and -separating processes. Components of multiple-piece products are then assembled into subassemblies and final assemblies with materials-combining processes, Figure 16.2.

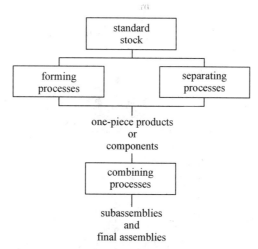

Figure 16.2 Secondary manufacturing companies use forming and separating processes to change standard stock into one-piece products or components of multiple-piece products. Combining processes are used to turn components of multiple piece products into subassemblies and final assemblies

16.3 Forming

Materials-forming processes convert standard stock (boards of wood, rolls or bars of steel, pellets of plastic, bags of ceramic cement or sand) into one-piece products or components of multiple-piece products. Forming processes can bring about internal or external form changes in metallic, ceramic, polymeric, or composite materials. External (outside) forms of materials can be changed through casting and molding and compressing and stretching processes. Internal forms of materials can be changed by using conditioning processes.

16.3.1 Casting and Molding

In casting and molding processes, industrial materials are melted, dissolved, or compounded into a liquid or semiliquid state. Then they are allowed to flow or are forged by pressure into a hollow cavity of a desired shape. Casting processes allow a liquid material to pour by gravity into a hollow cavity, Figure 16.3. Molding processes force liquid or semiliquid materials by pressure (injection) into a cavity. Metal materials like steel are melted in furnaces, ceramic materials like clay are dissolved in water, and plastic materials are com pounded (mixed).

The five basic steps in casting or molding any material are:
(1) preparing a pattern;
(2) preparing the mold;
(3) preparing (melting, dissolving, compounding) the material;
(4) introducing the material into the mold;
(5) removing the solid piece.

Perhaps the greatest advantage for casting and molding processes is that they provide a

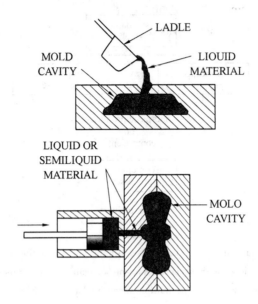

Figure 16.3　Casting occurs when liquid materials flow by gravity into a hollow cavity. Molding takes place when liquid or semiliquid materials are forced (injected) into a hollow cavity

direct route from industrial material to a semi-finished product at a relatively low cost.

16.3.2　Compressing and Stretching

A second major way that forming can be done is by compressing or stretching materials. Compressing processes squeeze and stretching processes pull materials into desired shapes.

Compressing and stretching can occur when any of four different kinds of forces are applied to materials. These forces are compression, tension, shear, and torsion. These four forces are usually applied to materials with compressing or stretching devices such as hammers, presses, rollers, or drawing machines.

When a small amount of force is applied to a material, it first changes shape (deforms) elastically. An elastic deformation is not permanent. The material will return to its original shape when the force is removed. Rubber is an example of a material with a great ability to deform elastically. A rubber band can be stretched around a bundle of envelopes. When the rubber band is removed, it returns to its original shape. Many other materials deform elastically, but this nonpermanent change in shape may not be visible to the human eye. Bricks and ceramic tiles are examples of such materials.

If the force on a material continues to increase, the elastic limit will eventually be exceeded. The shape change will then become permanent. This permanent shape change is called plastic deformation. Each type of material needs a different amount of force for deformation to occur. Therefore, compressing and stretching tools (hammers, presses, rollers) are designed to provide just the right amount of force to cause plastic deformation of a material into a desired shape.

Plastic deformation is easier when materials are soft. If the materials are not softened enough before compressing or stretching, they could be easily fractured and ruined before the final desired shape is formed. Water is added to clay and wood, while plastics, metals, and glass are heated before compressing or stretching.

There are seven basic ways to compress (squeeze) or stretch (pull) materials into, desired shapes. These are squeeze forming, rolling, extruding, drawing, stretch forming, spinning, and bending, Figure 16.4. Although specific process names may vary depending on the type of material (metallic, ceramic, polymeric, composite) that is being formed, the seven basic processes are the same.

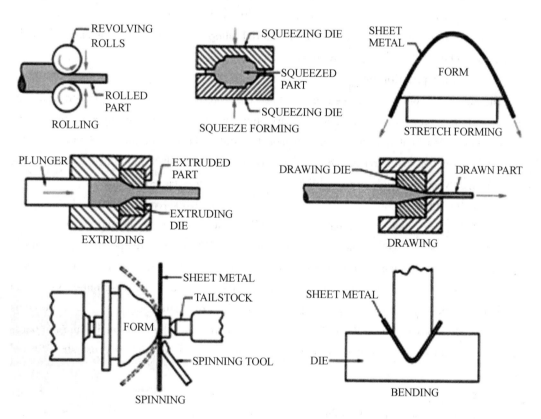

Figure 16.4　The seven basic ways to compress or stretch materials are squeeze forming, rolling, extruding, drawing, stretch forming, spinning, and bending

16.3.3　Conditioning

Conditioning processes are the third major category of forming processes. Materials are conditioned to change their internal structures on a cellular or atomic level. These internal changes usually are not visible. For example, a malleable piece of carbon steel can be heated until it is cherry red, then plunged in water. The steel does not look any different, but profound changes have happened to its internal structure to make it stronger. The material properties that are changed through conditioning processes include hardness, brittleness, toughness, elasticity, plasticity, ductility, strength, and moisture content.

There are three basic reasons why materials are conditioned. First, properties may need to be changed so that materials are easier to process. Hard materials are commonly softened with heat so that they are more easily compressed, stretched, sheared, machined, or welded into desired shapes. Second, materials may be conditioned to relieve internal stresses and strains that can build up as they are processed. High internal stresses and strains can cause a material to become hard, brittle, and easily fractured. A third reason for conditioning is to give materials some special property or properties desired for the final product.

There are three malor ways to condition materials-thermal conditioning, mechanical conditioning, and chemical conditioning processes.

1. Thermal Conditioning

Thermal conditioning processes use heat to change internal structures and properties of materials. Metallic and some ceramic materials are often thermally conditioned by a group of related processes called heat treating. Common heat-treating processes include hardening, tempering, and annealing. Clay-based ceramic materials are thermally conditioned by a process called firing. Wood materials are conditioned through a process called drying.

2. Hardening

Hardening takes place when material (iron alloys such as steel) is heated above a critical temperature and is then very rapidly cooled or quenched. The critical temperature is a point at which the material will change from one crystal structure to another. Certain characteristics that allow a material to become harder (such as fine grain size and an even distribution of carbon throughout all molecules) can only take place above the critical temperature. The material is rapidly cooled to lock or freeze these desirable characteristics into the material. While the hardening process causes a material to become harder, it also leads to high internal stresses and brittleness.

3. Tempering

A tempering process usually follows hardening. The material is heated to a predetermined point below the critical temperature and then is quenched. Tempering relieves internal stress, reduces brittleness, and provides toughness to materials. Low tempering temperatures result in less toughness. Higher temperatures lead to greater toughness.

4. Annealing

Annealing means to soften. Materials are heated to a point above their critical temperature. They are then allowed to cool slowly. The slow cooling permits the grain size and distribution of carbon to return to normal. The annealing process can be used to remove any effects of thermal hardening or tempering as well as hardening caused by compressing and stretching processes.

5. Drying

Drying is a conditioning process that removes moisture from materials. This process is usually associated with wood and some ceramic materials. The kiln drying process removes excess moisture from most furniture grade lumber. Construction grade lumber is usually air dried. Freshly cut (green) lumber has about an eighty-five percent moisture content. The kiln drying process slowly reduces the moisture content to about eight percent. As the moisture content is reduced, the hardness, strength, and stability (resistance to warping) of the lumber is increased.

Clay-based ceramic materials are usually dried in an oven after they have been formed to prepare them for a firing process. The firing process cooks the clay product at extremely high temperatures. The product becomes hard, brittle, and permanently solid.

6. Mechanical Conditioning

Mechanical conditioning processes use physical force to change materials properties. Materials may be compressed, stretched, flexed, or hit to cause internal changes in the structure of the material. Mechanical conditioning may also be called work hardening. Work hardening occurs when compressing or stretching processes are performed on cold metal materials. The physical force transferred to the material by these processes changes the size and spacing of internal metal grains. The grains become elongated in this rolling process, and the overall hardness of the material is increased.

7. Chemical Conditioning

Chemical conditioning processes cause chemical reactions to take place in materials. When this happens, the atoms inside a material rearrange themselves into a new chemical structure. The material that causes a chemical reaction to take place is called a catalyst. A catalyst material can be added to a thermosetting plastic material like auto body repair putty to cause it to quickly harden. The catalyst causes the thermosetting plastic to polymerize combine short molecular chains of atoms into longer, more stable chains. These changes cause the thermosetting plastic to become harder and much more stable than thermoplastic materials. Other chemical conditioning processes include leather tanning, as well as materials curing by exposure to gamma ray radiation.

16.4 Separating

Materials-separating processes are performed for the same basic reason as materials-forming processes-to convert standard stock into one-piece products or components of multiple-piece products. Separating processes differ from forming processes because they are used to cut down the size of materials by removing excess material. The material left over after all excess has been removed (separated) is the desired product or component. Forming processes result in products or components by changing the shape of the material without removing any material.

There are three basic ways to separate excess material from standard stock. The two commonly used groups of processes are shearing and chip removing. A third group represents special separating processes that are not used as often or are relatively new processes that do not fit into either the shearing or chip-removing categories. This third group of processes is simply called special separating processes.

16.4.1 Shearing

Shearing processes are used to cut excess material away from standard stock with no loss of material. Let's say that a twelve-inch-long piece of material was shear-separated into two six-inch-long pieces. If the two six-inch pieces were placed back together, they would measure the original twelve inches.

Excess material that is shear-separated is usually recycled or can be designed for use in other products. Therefore, shearing processes are very material-efficient because little if any stock is wasted.

Shearing processes generally use two opposing surfaces to produce a fracture or break in the material. A movable shearing tool is brought into contact with the material. The shearing tool must be harder than the material being cut. Pressure is then applied, and the tool penetrates the material. As penetration continues, the material fractures and the part are separated.

Shearing processes are usually performed with one of three major kinds of cutting tools: shears and blades, punches and dies, and rotary cutters.

1. Shears and Blades

Shears and blades are often used for trimming books and magazines, shearing metal and plastic sheets, shaving whiskers, trimming tree branches, cutting wood veneer, and cutting extruded ceramic bricks to length before firing.

2. Punches and Dies

Punches and dies are often used to produce holes in sheet materials. The size and shape of the separated material or the hole left behind is controlled by the shape of the punch and die. A paper hole puncher is a good example of how a punch and die works to shear-separate material.

3. Rotary Cutters

Rotary cutters produce round and irregular shapes. Revolving circular cutters as well as roller-razor blade combinations, are used to cut rolls of metallic, plastic, paper, fabric, and other thin materials to finished sizes.

16.4.2 Chip Removing

Chip-removing processes, like shearing processes, are used to cut excess material away from standard stock. Chip removal, unlike shearing, results in a loss or waste of material in the form of chips (small bits of the stock). If a twelve-inch-long piece of material were chip-separated in half by a saw with a 1/8 inch kerf, the result would be two $\frac{95}{16}$-inch-long pieces. If the two pieces were placed back together, they would measure $\frac{95}{8}$ inches long. A one-eighth-inch saw kerf of material would have been lost, it would have been turned into chips by the saw blade teeth.

Excess material that is chip-separated can be recycled or can be designed for use in other products; however, it is not a common practice to do so. Therefore, chip-removing processes are not as material-efficient as shearing processes.

Chip-removing processes separate materials by forming and removing chips. Chips are formed as a result of a cutting tool being fed into, along, or through a material. The tool compresses the material ahead of it. The material first changes shape elastically and then permanently deforms. Finally, the material fractures at a place called the shear plane, moves up and over the face of the cutting tool, and breaks away from the parent material in the form of a chip.

There are five basic ways to separate materials by chip removal. These are sawing, hole machining, turning, milling, and abrasive machining.

16.5 Combining

Materials-combining processes are used to finish one-piece products and to assemble components of multiple-piece products into sub-assemblies and final product assemblies. Materials can be combined by mixing, bonding, coating, and mechanical fastening.

16.5.1 Mixing

Mixing is a combining process used commonly in food, petrochemical, and drug manufacturing companies. Mixing processes evenly spread molecules of gas, droplets of liquid, or particles of solid materials throughout the mixture.

Dry (solid) materials are usually mixed by stirring, tumbling, mulling, and fluidizing. Correct portions of cement and sand are stirred with a shovel before water is added to make concrete. Entry postcards, balls, or other markers used in raffle, bingo, sweepstakes and lottery competitions are tumbled in drums so that all will have an equal chance of being selected. Mulling processes mix fine granular solid materials like metal casting sand and baking flour. The ingredients are placed in a circular tub. Power-driven wheels roll over the materials. This causes them to move sideways and mix together. Fluidizing occurs when air currents are blown through powdered ingredients, causing them to mix.

Liquid materials are typically mixed with pouring, shaking, blending, agitating, and vibrating processes. Thin liquids like water can be mixed by simply pouring them together. Thick liquids such as molasses or heavy-grade oils might need special vibrating, blending, or agitating processes to mix with other liquids. Liquid paint colors are mixed with a base liquid in a special vibrating machine at the local hardware or paint store. The food industry uses blending processes to make such products as breads, cakes, and ice cream in large quantities. Air can be bubbled through liquids to mix them in a process called pneumatic agitation.

Gases are mixed by a process called metering. Metering means to measure by volume. When gases are to be mixed, each gas is metered. The correct proportion (fraction) of each gas is then simply fed into a closed container. Since gas molecules move much faster than molecules of liquid and solid materials, the gases quickly mix themselves inside the container.

Many combinations of solids, liquids, and gases are also combined by mixing processes. Solid materials like salt or sugar can be mixed with water or other liquids by a process called dissolving. Mixtures that occur when solids are dissolved in liquids are called solutions. If the solid cannot be dissolved in a liquid, it becomes a suspension, which is evenly spread throughout the liquid. Sand and gravel are examples of materials that are suspended in a concrete mixture. Finally, gases and liquids are commonly mixed. Many soft drinks are carbonated by forcing carbon dioxide gas to dissolve into the liquid drink. When the bottle or can is opened, the gas bubbles rise to the surface.

16.5.2 Bonding

Bonding processes combine (assemble or join) solid materials together. The area of the material where the bond takes place is called a joint. The type of joint selected for use depends on the strength and appearance requirements of the product as well as the type of material that is to be bonded.

Bonding is a permanent combining process. Bonded joints are rigid so that parts that are bonded cannot move or be separated. Materials can be bonded by using either adhesive or cohesive processes.

Adhesive bonding processes use glues, cements, or some other adhesive to cause the joint areas of parts to stick together. The adhesive is applied in a thin layer at the joint area of the mating parts that are to be combined.

Adhesive bonding processes can be used to bond metallic, polymeric, and ceramic materials. Examples of adhesive bonding include gluing wood products, soldering copper pipes or components to printed circuit boards, cementing firebrick in kilns or furnaces, taping sheets of paper together, ironing patches onto clothing to cover holes, and pasting up wallpaper.

Cohesive bonding processes use heat, pressure, chemicals, or combinations of each to melt or soften parts at the joint. The molecules of the melted surfaces then move across the joint and cohere or mix together. This combining process is also referred to as fusion bonding.

Cohesive (fusion) bonding processes can be used to bond metallic, polymeric, and ceramic materials. Examples of cohesive bonding include many kinds of gas, electrical, and ultrasonic (vibration) welding processes that create heat. Also included are cold welding processes, which use tremendous amounts of pressure to force molecules to cross joint areas and cohere. Other examples of cohesive bonding include metal brazing (which differs from welding because filler materials that are very different than the base materials are used) and the use of special chemicals (such as ethylene dichloride) to soften and combine thermoplastic materials at joint areas.

16.5.3 Coating

Coating processes place one or more layers of one material on top of another material. Materials are coated to protect them from wearing out too fast, to decorate them so they will have an attractive appearance, to communicate information, to reduce noise, or to reflect heat or light. There are two major ways to coat one substance with another: (1) by applying a physical coating; (2) by converting (changing) the surface of a material.

Physical coating processes are used to apply a thin layer of coating material to the surface of another material by brushing, rolling, dipping, spraying plating, or printing. The coating material sticks to the surface of the part or product being coated.

Conversion coating is used to change chemically the surface of a material. It includes dyeing, oxide coating, or phosphate coating. Conversion coating processes do not add additional material to the outside surface of another material. The outside layer of the material is simply changed to improve appearance or wearability.

Phosphate coating treats metals in a chemical solution that causes the surface of the material to become a very good undercoating for additional physical coatings of paint or some other exterior finish. Another conversion coating process called anodizing stimulates aluminum or magnesium metals to "grow" a clear or colored protective oxide coating.

16.5.4 Mechanical Fastening

Mechanical fastening, perhaps the oldest process used to combine materials, has been traced to the early cave dwellers' use of vines to lace stones and sticks together to produce crude tools and weapons. Examples of mechanical fasteners include staples, nails, screws, rivets, pins, nuts, bolts, thread, keys, shafts, hinges, and clamps. The two major kinds of mechanical fasteners are nonthreaded fasteners and threaded fasteners.

Mechanical fastening processes rely on physical force to hold parts together. Nonthreaded mechanical fastening processes such as nailing, tacking, or pinning compress (squeeze) the material that they enter. The pressure created by the compressing of the material holds the fastener in place by friction. Other examples of nonthreaded mechanical fastening processes include sheet metal seaming; spring clamping; the sewing and weaving of textile materials, flat or wire strapping; and fastening with rope, cable, or chains.

Threaded mechanical fasteners are much better holding devices than nonthreaded fasteners. The surface areas of threaded fasteners such as screws, bolts, and nuts are much larger than the surface areas of nonthreaded fasteners. Larger surface areas allow greater frictional forces to develop, therefore, parts can be held much better. Threaded fasteners also provide the advantage of non-permanence. Products combined with threaded fasteners can be easily disassembled for repair or maintenance.

Words/Phrases and Expressions

form 成形
metallic 金属的
ceramic 陶瓷的
polymeric 聚合的
composite material 复合材料
plant raw material 植物原料
animal raw material 动物原料
slaughter 宰杀
domesticated animal 家畜
sheets of plywood 胶合板
particleboard 刨花板
granule 颗粒
pellet 颗粒
plastic resin 塑料树脂
cement 水泥
pig iron 生铁
steel 钢
aluminum 铝
copper 铜
ingot 铸锭
sheets 薄板
coil 卷材
melt 熔化
cast 铸造
roll 轧制
mold 成型
compress 压缩
stretch 拉伸
dissolve 溶解

compound 混合
liquid 液态
semiliquid 半液体
forge 锻造
hollow cavity 空心腔
furnace 熔炉
clay 黏土
compression 压缩
tension 拉伸
shear 剪切
torsion 扭转
rubber 橡胶
conditioning 调质处理
hardness 硬度
brittleness 脆性
toughness 韧性
elasticity 弹性
plasticity 可塑性
ductility 延展性
strength 强度
moisture content 水分含量
stress 应力
strain 应变
heat treating 热处理
hardening 淬火
tempering 回火
annealing 退火
firing 烧制
permanent 永久的
flex 弯曲
punch and die 冲头冲模
petrochemical 石化
stirring 搅拌
tumbling 翻滚
mulling 研磨
fluidizing 流化
agitating 搅动
bonding 粘接

ethylene dichloride 二氯乙烯
coating 涂层
phosphate 磷酸盐
non-permanence 非永久性

Chapter 17　Manufacturing Technology Today and Tomorrow

17.1　Introduction

Centuries ago, all products were made by hand. In fact, the word manufacture comes from the Latin manu factus, meaning "made by hand". Over the years, manufacturing has come to mean making products either by hand or by machine. More specifically manufacturing refers to the making of products in a factory. Today, most common products are made in factories using sophisticated machines. In our fast-paced society, we can seldom afford to spend the extra time and money needed to make products by hand.

Factories produce a large number of products quickly. Factories are well organized, Figure 17.1. Products are made by completing a series of production tasks in a systematic order. This systematic approach increases the efficiency of manufacturing. Efficiency means making good use of time, materials, and other resources. Efficiency lowers the cost of a product and often improves its quality.

Figure 17.1　Manufacturing in factories

Manufacturing begins with the original idea for the product. The steps that follow include research and development, planning for production, tooling up, production, and marketing and distribution.

17.2　Components of Manufacturing Systems

Every manufacturing system, no matter how simple or complex, includes basic components or parts. The parts shown include input, process, output, and feedback. "Input" is all the resources that go into making a product. "Process" includes all the actions carried out to run the company or make the product. "Output" is the completed product. "Feedback" includes such things as quality assurance and the reaction the customer has to the product. Keep the model in mind as you study this course. It will help you to understand how the parts fit together and how the system works.

17.2.1 Manufacturing Resources

Resources are all the materials, equipment, and labor necessary to manufacture a product. Manufacturing resources include:

①materials;
②tools;
③people;
④capital;
⑤energy.

Using these elements in different ways, a company can produce many types of finished products. Therefore, resources are often referred to as the "building blocks" of production.

1. Materials

Materials will be used to mean any items that can be processed to produce a finished product. Materials include two basic categories: natural materials and synthetic materials.

Natural materials, or raw materials, are those that are found in nature. These are the basic resources from which all other materials and products are made. Iron ore and wood are two examples of natural materials.

Materials that are made by people are called synthetic materials. Synthetic materials are not found in nature. They are the result of planned chemical reactions or other changes. These changes turn natural materials into entirely new materials. One common example of a synthetic material is plastic.

2. Energy

Without energy, factories could not operate. Energy powers the machinery that makes the factory run. It also makes possible transportation for supplies coming into the factory and products being sent to markets. Because energy has become so costly in recent years, every manufacturing company must take it into account.

17.2.2 Manufacturing Processes

Processes are those things done to change the size, shape, strength, or appearance of materials. There are three major types of processes in manufacturing: separating, forming, and combining. Separating is the process of removing or cutting materials. Forming is changing the shape of materials. Combining is the process of assembling materials or joining them together.

Many products manufactured require parts that themselves must be manufactured. For example, a radio contains such things as circuits and transistors. These electronic parts must first be manufactured before they can be used in a radio. Manufacturing processes can therefore be divided into two distinct categories: primary and secondary.

1. Primary Processes

Raw materials are seldom usable as they are found in nature. They must be processed before they can be used to make products. Procedures that prepare raw materials for use in manufacturing are called primary processes.

Sometimes primary processes are necessary to separate unwanted or impure materials from desirable materials. For example, crude oil has little value in its natural form. It must be processed at a refinery to separate it into such materials as motor oil, gasoline, and diesel fuel. Other natural resources must be shaped or cut into parts that can be used to make products. For example, an oak tree must be cut up into lumber, dried, and planed (smoothed) before it can be used to make furniture.

Products that are made specifically to be used in making other, more complicated products are called producer products. Bicycle handlebars are an example. Handlebars are produced specifically to be installed on a bicycle. If production of bicycles were to suddenly stop, there probably would be no other uses for handlebars.

2. Secondary Processes

Procedures in which prepared materials are used to manufacture more complex items are called secondary processes. Most secondary processes result in products that are eventually bought and used by consumers. These are called consumer products. Radios and bicycles are examples of consumer products.

Consumer products can be further categorized according to their durability or toughness. Generally, durable products are those that are designed to last at least three years. Examples of durable products are automobiles, lawn mowers, airplanes, and machines. Non-durable products are those that are designed to last only a short time. Examples of non-durable products are pencils and flashlight batteries.

17.3 Manufacturing and Technology

Manufacturing is a system with many components. Manufacturing is also a component of a larger system of human activities called "technology". Technology is knowing and doing things to extend human abilities and potential. For example, a calculator extends our ability to solve math problems when we know how to use it. Bicycles extend our ability to travel to school and back. Both the calculator and bicycle become technology when used by humans.

Manufacturing is technology because it allows humans to make products that extend their abilities and potential. Of course, manufacturing relies on several other technologies to make quality products in an efficient manner. For example, the manufacture of products requires energy/power technology to operate machines. It requires transportation technology to get the raw materials to the factories and to move the finished products to consumers. Manufacturing also uses communication technology to inform workers how to manufacture products and tell consumers about them.

Likewise, the generation and transmission of electricity (energy) relies on manufacturing to make generators, transformers, and wires. Transportation systems rely on manufacturing to produce the vehicles and construction to make roads, docks, and airports.

17.4 Changes in Manufacturing

Products and production methods are constantly changing. Advances in technology improve the tools, materials, and processes used. As a result, today we can make better products more efficiently than ever before. For example, computers are now used to help manufacture some products.

They help design the products as well as run the machines and assembly lines that produce them.

To stay in business, companies must appeal to consumers. They are always trying to produce new and better products to meet consumer demand. For example, if gasoline prices increase, consumers may want automobiles that use less fuel. Automobile manufacturers respond to consumer wishes by creating more fuel-efficient automobiles.

17.5 Timely Robots

Until recently, robots could only perform tasks that did not require precise movements. Even when controlled by a computer, robot motions were not very accurate. However, Japanese scientists developed extremely precise robots to perform delicate operations. In one factory, these robots are now being used to assemble watches.

These watch-making robots, like the ones shown here, were developed by Seiko/Epson. They are programmed to pick up a tiny watch part and place it in exactly the right position. Once the robots have been programmed, they can repeat their movements almost exactly. Their accuracy has improved a lot.

A line of robots puts the products together one step at a time. Each performs a simple task, such as picking up a gear and dropping it into place. Each operation takes only less seconds. Together, this amazing team of robots can have high productivity.

Scientists are exploring many other uses for precise robots. As robots become more advanced, they will be used to make even more complex products. Someday it may be unusual for any product to be built by human hands.

Words/Phrases and Expressions

building block 基石
natural material 天然材料
synthetic material 合成材料
raw material 原材料
impure 不纯的

Chapter 18　Intelligent Manufacturing and Case Study

Intelligent technologies are leading to the next wave of industrial revolution in manufacturing. Intelligent manufacturing is a general concept that is under continuous development. It can be categorized into three basic paradigms: digital manufacturing, digital-networked manufacturing, and new-generation intelligent manufacturing. New-generation intelligent manufacturing represents an in-depth integration of new-generation artificial intelligence (AI) technology and advanced manufacturing technology. It runs through every link in the full life-cycle of design, production, product, and service. The concept also relates to the optimization and integration of corresponding systems; the continuous improvement of enterprises' product quality, performance, and service levels; and reduction in resources consumption. New-generation intelligent manufacturing acts as the core driving force of the new industrial revolution and will continue to be the main pathway for the transformation and upgrading of the manufacturing industry in the decades to come.

The three major novel manufacturing concepts and technologies emerge in the environment of the fourth industrial revolution, also known as Industry 4.0 (Figure 18.1), are intelligent manufacturing, IoT-enabled manufacturing, and cloud manufacturing.

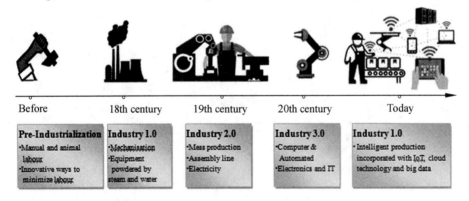

Figure 18.1　The evolution of the industrial revolution

(1) Intelligent manufacturing (also known as smart manufacturing) is a new manufacturing model based on intelligent science and technology that greatly upgrades the design, production, management, and integration of the whole life cycle of a typical product. The entire product life cycle can be facilitated using various smart sensors, adaptive decision-making models, advanced materials, intelligent devices, and data analytics. AI technology facilitates the development of new models, new means, and new forms, system architecture, and technology systems in the domain of intelligent manufacturing.

①New models. Internet-based, service-oriented, collaborative, customizable, flexible, and socialized intelligent manufacturing system that is used to facilitate production and provide services to users.

②New means. Human-machine integrated smart manufacturing systems featuring digitalization, Internet of Things, virtualization, service, collaboration, customization, flexibility, and intelligence.

③New form. Intelligent manufacturing ecology with the characteristics of ubiquitous interconnection, data-drivenness, cross-border integration, autonomous intelligence, and mass innovation.

The deep integration of the application of these models, means, and forms will ultimately form an ecosystem of intelligent manufacturing.

Based on the understanding of existing theories and practices, we conclude that intelligent manufacturing consists of intelligent products, intelligent production, and intelligent services. As shown in Figure 18.2, the basic framework of intelligent manufacturing and the interrelationship of its constituent elements are described. Intelligent products include intelligent facilities (i.e., sensors, data storage equipment, software, etc.), physical entities (i.e., parts, machinery, etc.), and networking components (i.e., interfaces, wired and wireless network protocols, etc.). Intelligent production is based on intelligent manufacturing system as the core, and intelligent factory as the carrier. It forms a complex manufacturing network through vertical integration of internal business and horizontal integration between enterprise value chains, realizing real-time management and optimization of product life cycle. Intelligent services conduct the management, analysis, mining of data generated from intelligent devices, and intelligent production process for intelligent decision-making.

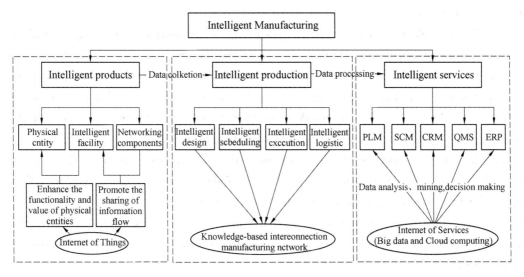

Figure 18.2 The basic architecture of intelligent manufacturing

(2) IoT (Internet of Things) -enabled manufacturing refers to an advanced principle in which typical production resources are converted into smart manufacturing objects (SMOs) that are able to sense, interconnect, and interact with each other to automatically and adaptively carry out manufacturing logics. Within IoT-enabled manufacturing environments, human-to-human, human-to-machine, and machine-to-machine connections are realized for intelligent perception. IoT-enabled manufacturing features real-time data collection and sharing among various manufacturing resources such as radio frequency identification (RFID), wireless communication standards, the visibility and traceability of various manufacturing operations.

The major components of the IoT are as follows.

①Sensing systems. Numerous heterogeneous sensing devices are attached through the Internet to give real-time data continuously.

②Outer gateway processors. The computing servers comprise application servers, edge servers, smart switches and smart routers, which are energy efficient and offer reduced latency. The outer gateway processors help to lower band-width utilization by reducing and filtering data at the edge of the internet.

③Inner gateway processors. These are composed of the micro-clouds, and cloudlet servers exist in the industrial wide area networks (WANs).

④Outer central processors. These are the computing servers beneficial in data filtration and data reduction and used for data processing, routing at inner central processors.

⑤Inner central processors. These comprise clusters, grids, clouds, and multi-cloud systems and stay far from the sensing system.

(3) Cyber-physical system (CPS) is defined as a technique to interconnect, manage and interact between physical devices and computational applications firmly integrated with the internet and its users.

CPS is a mechanism through which physical objects and software are closely intertwined, enabling different components to interact with each other in a myriad of ways to exchange information. In CPSs, physical and software components are working at diverse spatial, behavioural, and temporal scales and interrelating with each other in a lot of ways that change with the situation. A CPS-enabled system, unlike a traditional embedded system, contains networked interactions that are designed and developed with physical input and output, along with their cyber twined services such as control algorithms and computational capacities. Thus, a large number of sensors play important roles in a CPS. Today's manufacturing world is fully equipped with sensors by which heterogeneous data can be collected for analytics and decision-making. The automation can be done in a collaborative way with the development and employment of apparatuses and methods accompanied by the computational and data communication capabilities.

In the context of manufacturing, Cyber Manufacturing Systems (CMS) and Industrial Internet of Things (IIoT) denote the respective industrial counterparts of CPS and IoT. CMS or

Cyber-Physical Production Systems (CPPS) are therefore advanced mechatronic production systems that gain their intelligence by their connectivity to the IIoT. Therefore, CMS cannot be considered without IIoT and vice versa.

Overall, CMS and IIoT are not individual technologies with a closed theory framework, but rather an interdisciplinary blend from the domains of production, computer science, mechatronics, communication technology and ergonomics, see Figure 18.3.

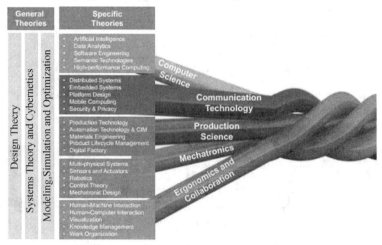

Figure 18.3 Theoretical foundations of CMS and IIoT

(4) Cloud manufacturing refers to an advanced manufacturing model under the support of cloud computing, the IoT, virtualization, and service-oriented technologies, which transforms manufacturing resources into services that can be comprehensively shared and circulated.

The clouds with infrastructure management capability are a huge lake of simply usable and reachable virtualized resources like different development platforms, applications, services, hardware which can be reconfigured dynamically to enable self-service, economies of scale and optimum resource utilization. Disseminated resources are captured into cloud services and managed in a centralized way in cloud manufacturing. Clients can take cloud services consistent with their necessities. The users of the cloud can request services of all stages of a product life cycle like product design, manufacturing, testing, management, etc. . Therefore, it is usually regarded as a parallel, networked, and intelligent manufacturing system (the "manufacturing cloud") where production resources and capacities can be intelligently managed. Figure 18.4 describes the working interfaces of cloud computing in intelligence manufacturing.

(5) Big Data Analytics (BDA), with an aggressive push toward the Internet and IoT technologies, data is becoming more and more accessible and ubiquitous in many industries, resulting in the issue of big data. Big data typically stems from various channels. For organizations and manufacturers with an abundance of operational and shop-floor data, advanced analytics techniques are critical for uncovering hidden patterns, unknown correlations, market trends, customer preferences, and other useful business information.

(6) Information and Communications Technology (ICT) refers to an extended IT that high-

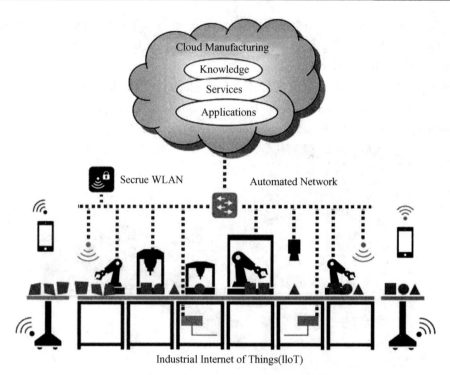

Figure 18.4 Cloud manufacturing and intelligence manufacturing

lights unified communications and the integration of telecommunications, as well as other technologies that are able to store, transmit, and manipulate data or information. ICT covers a wide range of computer science and signal-processing techniques such as wireless systems, enterprise middleware, and audio-visual systems. It focuses on information transferring through various electronic media such as wired or wireless communication standards.

During the upgrading of manufacturing capabilities, a group of Chinese firms with a good foundation in the digital manufacturing paradigm was able to successfully transition to the smart manufacturing paradigm, thereby becoming a demonstration project for "internet + manufacturing" in China. SANY Heavy Industry Co., Ltd. (hereafter referred as SANY) is one of these firms.

Case:

SANY was founded in 1994 and produces concrete equipment, excavators, cranes, and so forth. It is one of the world's leading providers of engineering equipment by far, as well as being the top provider of concrete-pump cars in the world. SANY was the first firm in China's engineering equipment sector to develop and adopt smart manufacturing technologies, which significantly improved the quality of its products. Following the digitalization of its production, SANY actively built a global IoT system and a big data platform to provide services such as PdM and IoT financial services, all of which have contributed to SANY's success.

SANY has made the digitalization of production a top priority in its strategy since the firm was founded. SANY has a firm-level strategy to control its digitalization process, which ensures that a good foundation is built before moving toward the next target. Between 1994 and 2004, SANY digitalized its key designing process and management system, and gradually integrated digitalized management into its daily routines.

With SANY's expansion, independent business management modules no longer met its need for a highly integrated management system. Thus, SANY spent a decade on using network technologies to link all independent modules starting in 2004. During this decade, a first-generation interconnected management system was built to share data and synchronize tasks between the design module and management module. Later on, SANY built a global operational management system to link and optimize the functionality of the subsystems built previously; this allowed it to start a new business model named "the internet of manufacturing". After Internet of Vehicles(IoV) technologies were introduced, SANY's management system extended toward the consumer end, forming a vertically integrated system from the customer end to the production end. The global platform also horizontally integrates the management of domestic and overseas business units, including SANY's marketing, sales, and after-sale services. With this highly integrated global platform, SANY built its market-analyzing system to further improve its responsiveness to the international market. In addition, it built a global collaborative R&D platform using network-enabled virtue reality (VR) and simulation technologies.

In 2015, manufacturing became one of China's critical national strategies, and Chinese firms actively responded to this strategy to conduct intelligent manufacturing upgrading. Since 2018, SANY has begun to explore the applications of AI in its products, production and services, thus taking the first step toward the new-generation intelligent manufacturing paradigm. It has developed several unmanned heavy machineries, including excavators and cranes, which can be remotely controlled with precision. Based on IoV technologies, SANY monitors and provides diagnostics for its products in real time. All data collected are used to provide high-value-adding services such as PdM and IoT finance. Table 18.1 shows the timeline of SANY's implementation of intelligent manufacturing technologies.

Table 18.1 Timeline of SANY's adoption of intelligent manufacturing technologies

Year	Progress in developing intelligent manufacturing technologies
1994-2004	CAD; system applications and products(SAP); accounting system; data center; OA; global video conference system
2004-2014	Three-dimensional design; product data management(PDM); product lifecycle management (PLM); global ERP; ERP central component (ECC); CRM; SCM; e-human resource (eHR); accounting analysis system; MES; largest digital plant in Asia; "internet + manufacturing"; e-commerce
2015	Horizontal integration of value chains; market analysis system; vertical connection of production process; product design based on VR and simulation
2018	Unmanned machineries; remote maintenance platform; financial service based on IoT and big data

Words and Expressions

intelligent manufacturing mechanism 智能制造
IoT(Internet of Things)-enabled manufacturing 物联网制造
cloud manufacturing 云制造
big data 大数据
smart manufacturing 智能制造
radio frequency identification (RFID) 射频识别

Part Ⅱ Modern Food Processing Equipment

Chapter 19 Modern Common Food Processing Technology

Food processing technique simplifies the set of methods and techniques used to transform raw ingredients into food or to transform food into other forms for consumption by humans or animals either in the home or by the food processing industry. The common food processes include: baking, bread making, dairy product manufacture, drying, thermal processing, fermentation, freezing, infusion, juicing, malting, milling, parboiling, peeling, peeling and cooking, storage, storage and milling, washing and winemaking.

1. Baking

Baking is the technique of prolonged cooking of food by dry heat normally in an oven. It is primarily used for the preparation of bread, cakes, pastries and pies, tarts, and quiches. It is also used for the preparation of baked potatoes, baked apples, baked beans.

2. Bread Making

Commercially produced bread is an important component of every day diet in many countries. During the bread making process, flour is subjected to biological (fermentation) and physical (baking) transformation.

3. Dairy Product Manufacture

Milk and milk products form a main constituent of the daily diet. Butter, cheese and yoghurt are the popular dairy products. Hence, it is important to study the effect of milk products manufacture and their metabolites.

Butter is made from milk. In cheese making, the heating and salting stages are important. Skim milk was recombined with butter oil that had been fortified. Hence, consumption of heat treated milk and dairy products may be safer than liquid milk.

4. Drying

Drying is the oldest method of preserving food. As compared with other methods, drying is quite simple. Food can be dried in several ways, for example, by the sun or in an oven or a food dryer can also be used.

5. Fermentation

Fermentation is a simple process during which the enzymes hydrolyze most of the proteins to amino acids and low molecular weight peptides. Starch is partially converted to simple sugars which are fermented primarily to lactic acid, alcohol and carbon dioxide.

6. Freezing

Freezing food is a common method of food preservation which slows both food decay and most chemical reactions. The activity of microorganisms is lowered at low temperatures.

7. Infusion

Tea and coffee are popular beverages throughout the world. A cup of tea or coffee that cheers can also be an important culture of human life. The dried (made) tea is subjected to an infusion process prior to human consumption.

8. Juicing

Commercial juice extraction operations typically use whole fruit. The juice processing is processed by washing, peeling, enucleation, crushing and stirring.

9. Malting

Malting is a process applied to cereal grains, it is a combination of two processes, notably germination and the kiln-drying process.

10. Milling

The mechanical method is used to overcome the cohesive force inside the solid material so that the large-sized material particles are turned into small-sized solids.

11. Parboiling

Parboiling means precooking of rice within the husk. Parboiling involves first hydrating paddy followed by heating to cook the rice and finally drying of the rice.

12. Peeling

Peeling is an important step in the processing of most fruits and vegetables. Chemical peeling (mostly lye peeling), mechanical peeling (mainly abrasion peeling), steam peeling and freeze peeling are conventional methods for peeling in the processing of fruits and vegetables.

13. Peeling and Cooking

Generally vegetables are consumed after peeling which can be followed by cooking.

14. Storage

Food preservation is to prevent food from spoiling and deteriorating, and to extend the period of consumption. It is a processing and treatment measure for long-term preservation of food. Commonly it includes physical and chemical preservation and storage methods, such as cryopreservation, high temperature preservation, dehydration preservation, improvement of osmotic pressure, improvement of hydrogen ion concentration, irradiation preservation, air isolation, addition of preservatives and antioxidants.

15. Storage and Milling

Stored grains are milled prior to their usage in various forms so the combined effect of storage and milling assumes considerable significance.

16. Thermal Processing

Application of heat to food commodities is commonly done through ordinary cooking, pressure cooking, microwave cooking, frying, sterilization and canning.

(1) Canning.

This commercial process in its various forms combines elements of washing, peeling, juicing, cooking and concentration.

(2) Cooking.

Cooking is the act of preparing food for eating by the application of heat. It encompasses a vast range of methods depending on the customs and traditions, availability and the afford ability of the resources.

17. Washing

Washing is the most common form of processing which is a preliminary step in both house hold and commercial preparation. Loosely held useless residues are removed with reasonable efficiency by varied types of washing processes.

18. Washing and Cooking

Washing the apples followed by cooking, steaming, drying, peeling and juicing. It is evident that the effectiveness of washing depends upon four factors. Firstly, the locations of the attachments where by the surface attachments are amenable to simple washing. Secondly, the age of the attachment is an important parameter since with increasing time the attachment tends to move into cuticular waxes or deeper layers so the amount of residue that can be removed by washing declines. Thirdly, the water solubility of attachment reflects not only their higher solubility in the wash but also the reduced propensity moved into waxy layers. Lastly, the temperature and type of wash also affect residue removal. Hot washing and blanching are more effective as compared to cold washing and the effectiveness may be improved further by detergent.

19. Wine Making

It is produced by fermentation. Various kinds of food are made by fermentation, such as wine, vinegar, soy sauce, etc.

19.1 Mixing Techniques

Mixing is widely used in the food processing industry to ensure uniform product quality. These mixers can be used for a wide variety of doughs ranging from sugar wafer batters to extremely tough or dry doughs. For use in blending premixes or preparing adjuncts a considerably wider choice is available. Ribbon blenders, twin-cone, or V-shell mixers are suitable for preparing premixes of drying ingredients. A large number of different designs for liquid blending are available, most of them being based on propeller-in-tank concepts. For mixing air in batters, devices such as the Whizzolator and the Morton pressure whisks can be used. There is increasing need for setting scientific criteria to optimize the design and performance of mixers.

There are several possible methods for classifying mixers. Horizontal mixer is a typical kind, which includes fixed bowl or tilting bowl. They are almost essential when gluten development is desired since vertical mixers are too inefficient and slow in this operation and spindle mixers do not have the right kind of action. Several American manufacturers construct horizontal mixers suitable for mixing cookie doughs or cracker sponges and doughs. All of this equipment has in common a horizontal mixing bowl, U-shaped in cross section, mounted in a heavy rigid frame enclosing the drive motor and transmission.

There are two methods used for discharging dough. In some models, the bowl can be tilted so that the top is brought to a forward facing position. Figure 19.1 illustrates a mixer. There are various forms of agitators (Figure 19.2).

Figure 19.1　A horizontal mixer

Figure 19.2　Agitators of forms

19.2　Dehydration

Dehydration is probably the oldest known method of food preservation. The drying of fruits in the sun and the smoking of fish and meat are both well-known processes that originated in antiquity.

The drying of food products is not used as extensively as it could be for a number of reasons, not the least of which is a perceived loss of quality. A dried food product offers the advantage of decreased weight, which has the potential for savings in the cost of transporting the product. However, there is often a decrease in the quality of the dried product because most of the conventional techniques use high temperatures during the drying process. The consumers demand only the highest quality finished product: one with little or no loss in sensory characteristics and with the advantage of added convenience. As many of the new techniques use lower temperatures and/or decreased drying times, they should be considered for use by the food industry. In addition, there is potential for the development of unique products using these unconventional techniques.

19.2.1 Current Dehydration Techniques

Most drying processes occur in two distinct stages. The first stage is the constant rate stage. Moisture removal occurs at a constant rate because the internal moisture of the product is transported to the surface at the same rate as that at which evaporation occurs. The second stage is the falling rate stage. This may actually occur at several different rates as the product changes phases. When a critical moisture level is reached, drying occurs at a decreasing rate. This critical point is a unique characteristic of each product.

Current common Dehydration Techniques include: solar (open-air) drying, smoking, convection drying, drum drying, spray drying, fluidized-bed drying, freeze drying, and so on. Many other drying methods have been designed for specific products, for example explosive puffing or osmotic drying for solute-infused products.

19.2.2 Novel Dehydration Techniques

1. Microwave Drying and Dielectric Drying

These techniques use the electromagnetic wavelength spectrum as a form of energy, which interacts with the materials, thus generating heat and increasing the drying rate dramatically. Dielectric drying uses frequencies in the range 1–100 MHz, whereas microwave drying uses frequencies in the range 300–300 000 MHz. Dielectric drying has been used to dry crackers and cookies as well as some cereal products. Microwave drying is currently being used to dry pasta products.

2. Microwave-Augmented Freeze Drying

Conventional freeze drying can be speeded up by using a volumetric heating mode, such as microwaves. By using microwave energy to augment the convection heating of freeze drying, drying rates can be increased by as much as an order of magnitude. As with any microwave processing procedure, a major drawback is the non-uniformity of the energy within the chamber. However, this problem can be partially offset by the use of waveguides and a rotating tray.

3. Centrifugal Fluidized-Bed Drying

The centrifugal fluidized-bed dryer works on the same principle as the conventional fluidized-bed dryer except that a rotating chamber is used. A schematic diagram of a centrifugal fluidized-bed dryer is shown in Figure 19.3. The product to be dried is loaded into the chamber, which is then closed. Hot air is introduced into the bottom of the chamber, which is rotated at relatively high speed. The rotational speed and the flow rate of the air must be balanced to ensure that fluidization is achieved. If fluidization does not occur, the particles will tend to adhere to the walls of the chamber when it is rotating. Some drying will still occur, but the process will not be as efficient. The same restrictions regarding the particle size and shape as with the non-rotating fluidized-bed dryer apply here. By using the centrifugal force to counter the increasing air flow, thus assuring fluidization, the drying rate is significantly increased.

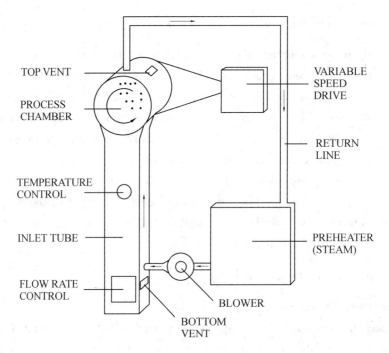

Figure 19.3 Schematic diagram of a centrifugal fluidized-bed dryer

4. Ball Drying

A schematic diagram of a typical commercial ball-drying system is shown in Figure 19.4. The material to be dried is added to the top of the drying chamber through a screw conveyor. Heated air is also added continuously to the chamber. The material within the drying chamber comes into direct contact with heated balls made from ceramic or other heat-conductive material. Drying occurs primarily by conduction. When the product arrives at the bottom of the chamber, it is separated from the balls and collected. Except for temperature, the most important variable to control is that of rotational speed. Relatively small particles without excessive sugar content such as vegetable pieces must be used.

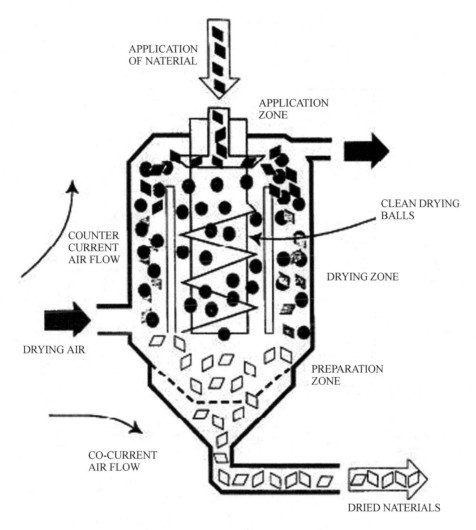

Figure 19.4 Schematic diagram of a ball dryer

5. Ultrasonic Drying of Liquids

It has been reported that ultrasonic energy can be used to remove liquid water from solutions of food particles. To use this process, the liquid is atomized to produce small-diameter droplets, first by a nozzle and then by further cavitation using ultrasonic energy within a drying chamber. The particles are then subjected to heating to remove the water, and the dried residue collected. The technique can greatly increase the evaporation rate of the water (with drying sometimes occurring in seconds). The procedure works best with low-fat solutions because oily or fatty foods do not dry effectively in an aerosol, as shown in Figure 19.5.

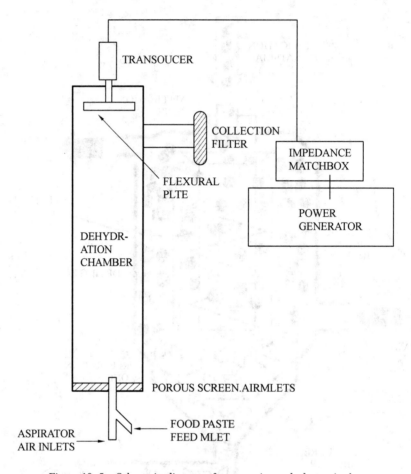

Figure 19.5　Schematic diagram of an experimental ultrasonic dryer

19.2.3　Conclusions

　　A number of novel drying techniques were used as alternatives to the morewell-known methods for moisture reduction in foods. Before a particular process is selected, consideration should be given to many factors, including: the type of product to be dried, the finished product desired, the product's susceptibility to heat and the cost of processing. There is no one "best" technique for all products.

　　In the future, it is probable that other novel drying techniques will be developed and become available for specialized purposes. Furthermore, current techniques will also probably be further refined to make them more economical and will also be explored for use with other food products.

19.3 Hygienic Design of Centrifugal Pumps for the Food Processing Industries

A pump is a machine that transports or pressurizes fluid. It transfers the mechanical energy of the prime mover or other external energy to the liquid to increase the liquid energy. The pump is mainly used to transport liquids such as water, oil, acid-base liquid, emulsion, suspension emulsion and liquid metal, as well as liquids, gas mixtures and liquids containing suspended solids. Pumps can usually be divided into three types: positive displacement pumps, power pumps and other types of pumps according to working principles. In addition to classification by working principle, it can also be classified and named by other methods.

Processing foodstuffs and beverages requires strict hygienic conditions throughout, and this presents some particular challenges to the pump manufacturer. Different aspects must be taken into consideration, including the design of the pumps, selection of materials, and legislation and regulations from the EEC, the Food and Drug Administration and other bodies product safety, consumer protection, reducing cleaning costs with regard to economic and ecological aspects, and legal requirements related to ease of cleaning (as contained in the EU Machinery Directive and other legislation) determine the requirements profile for the proper hygienic design of components. These components include pumps, which in the case of contaminants can spread the seeds of contamination throughout all parts of a facility. Consequently, the hygienic design and cleanability of all components is an indispensable requirement for all processes in beverage and food processing engineering.

The following parameters must be taken into account in the hygienic design of centrifugal pumps, and of course they are typically equally applicable to other components: construction; materials; legislation, directives and regulations.

19.3.1 Construction

The most important criterion for the cleanability of centrifugal pumps is primarily a type of construction that is easily cleaned. As trite as it may sound, the prerequisite for hygienic design is usually a simple, uncomplicated construction. In particular, with regard to construction, this means that all components that come in contact with the product, especially sealing surfaces, must be free of gaps, the contact pressure of sealing materials must be well defined using metal to metal contact and a hygienic, certified shaft passage or seal must be used. Another major consideration is that the circulation space should be free of dead ends. A further indispensable construction criterion is the use of aseptic connections.

19.3.2 Materials

The selection of materials and their surfaces with regard to cleaning and corrosion characteristics plays an important role in the quality and safety of the entire process. This means that the requirements profile includes the specification of a defined surface roughness, free of fissures and scratches, as well as the use of non-toxic, taste-neutral and corrosion-resistant materials that are approved for use with foodstuffs or are FDA compliant.

19.4 High Pressure Processing in Food Industry

The increased consumers' interest in high quality foods with fresh-like sensory and additive free attributes led to the development of non-thermal food processing technologies as alterative to conventionally heat treatments. The investigated inactivation technologies are, for example, high pressure processing (HPP), pulsed electrical fields, UV decontamination, high power ultrasound, and oscillating magnetic fields.

In food industry, as the most successfully commercialized non-thermal processing technology, high pressure processing (HPP) technology eliminates food pathogens at room temperature and extends the shelf life of foods circulated through the cold chain. These processes maintain the organoleptic properties and nutritional value of the foods, which is not possible using traditional thermal pasteurization. The high pressure processing is the technology by which a product is treated to a high level of hydrostatic pressure (>200 MPa) during a specific period of time. The number of foods suitable to undergo high pressure processing is very large, including a wide variety of meat products (boiled ham, cured ham...), fish, ready-to-serve meals, as well as the majority of fruit, vegetables and juices.

Despite the high price and high barriers to investment, the specialized original equipment manufacturer service sector has been gradually increasing. HPP technology also can be combined with existing food trends, as organic food, health food or clean label to boost the development in the food market. HPP technology has been widely used in the production of meat products, dairy products, aquatic products, vegetable and fruit products, and various beverage products.

19.4.1 Description of the High Pressure Equipment

HPP devices are divided into two types, the horizontal and vertical types. Most devices used in commercial applications are the horizontal type to facilitate the loading and unloading of containers in the production line.

To carry out the high pressure processing, it is necessary to design high-pressure equipment that has a sufficient capacity to work cyclically on a production line. Nowadays, this processing is achieved by subjecting the product to high pressures inside huge vessels capable of pasteurising large volumes of product. Water is the medium that transmits the pressure, which means that, before being processed, the products must be packaged in a flexible container capable of resisting the small variations of volume that take place during the operation.

To reach the processing pressure, p_w = 500 MPa in the present vessel, and achieve a life span of the high-pressure machine that makes it cost-effective, more than 100 000 work cycles must be carried out. For this, the vessel and some of the machine parts can be designed using the autofrettage technique. This autofrettage technique consists of subjecting the vessel to an over-pressure which locks plastic strain in an internal core, so that when the vessel is unloaded some residual compression stresses are generated in the previously plastic core. This way, when the vessel is subjected to its interior work pressure, the mean tangential stresses in the inside decrease considerably and, as a consequence, the fatigue resistance of the vessel is significantly improved.

In recent years, many studies have been carried out in order to determine the state of stress that appears after the autofrettage process. In plasticity, most of the steels undergo a phenomenon known as the Bauschinger effect, which consists of a reduction of the yield stress in compression as a result of a previous tension yielding. This effect reduces the theoretical residual compression stresses that are generated during the autofrettage process and it must therefore be taken into account during the fatigue design of the vessel.

19.4.2 Applications of High Pressure on Foods

HPP technology eliminates bacteria in packaged food and can be used in this capacity as one of the antibacterial factors in hurdle technology, in combination with various processing technologies, without changing the original process. HPP technology maintains organoleptic properties of vegetable and fruit products, eliminates micro-organisms and increase shelf-life. Nutritional values of food were well retained.

In the future, HPP will become an important pre-treatment step in the research and development of components with healthcare functions for improving the production performance of functional components in the food and pharmaceutical industries.

19.4.3 Future Trends

HPP technology extends the storage life, maintains the flavors and nutritional value. In addition, it extends the shelf life. Its nonthermal processing nature makes HPP technology a preferred production choice for maintaining food quality. However, this technology has the following disadvantages.

(1) Most HPP products must be stored and transported under refrigeration because pressure treatment at ambient or chilled temperature is not sufficient to some inactivate spores of harmful pathogens.

(2) This technology is not applicable to several food types, such as flour and powdery flavors with low water content or products containing a large number of air bubbles because HPP requires the use of water as a pressure transfer medium and products containing air bubbles will be deformed under pressure.

(3) The packaging material used in HPP must have a compressibility of at least 15%, so only plastic packaging materials are suitable for HPP.

Thus, an important issue for the future growth of HPP technology in the food sector is the establishment of relevant laws and regulations. Clearly specified process conditions will ensure microbial safety, product quality, and compliance with food hygienic safety laws and regulations in countries where the manufacturers are located. For example, high-quality organic food raw materials, fresh local foods with short food miles, and functional health foods are future development trends in thefield of HPP.

19.5 Heat Transfer Mechanisms

Heat may be transferred from one location to another by convection, conduction or radiation.

19.5.1 Convection, Conduction and Radiation

Convection is the transfer of heat from one part to another within a gas or liquid by the gross physical mixing of one part of the fluid with another.

Conduction is the transmittal of heat from one part to another part of the same body, or from one body to another in physical contact with it, there being no appreciable displacement of the particles of the bodies.

Radiation transfer of energy by electromagnetic radiation is a significant factor. These radiations are not in themselves heat, but are converted into heat through absorption by, and interaction, with, absorbing molecules. Light, which is an electromagnetic radiation, is all extremely poor means of transferring heat. Invisible infrared radiation is quite effective.

19.5.2 Heat Exchangers

There are various kinds of heat exchangers, including tube heat exchanger, plate heat exchanger and so on.

1. Heat Transfer in a Plate Heat Exchanger

Plate heat exchangers (PHEs) are extensively used for heating, cooling and heat-regeneration applications in the chemical, food, and pharmaceutical industries because of their high thermal efficiency, flexibility and ease of sanitation. The PHE consists of a pack of metal plates pressed together into a frame. A sequence of thin channels is formed between the plates and the flow distribution of the hot and cold streams is defined by the plate perforations and the gasket designs; thus a large number of configurations are possible. For example, the plate pasteurizer has stainless steel flat plates with silicon gaskets. Fluid was distilled water on hot and cold sides of the PHE.

The equations for heat and momentum transfer are solved (Figure 19.6). Figure 19.7 shows the expanded CFD representation of the two PHE configurations without the plates.

Figure 19.6 Fluid domain for series flow arrangement of PHE (a) The plate pack of the PHE; (b) The CFD representation of the fluid domain for series flow arrangement showing a mesh detail

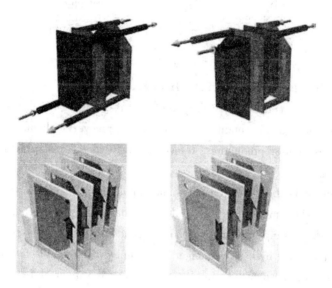

Figure 19.7 Expanded CFD representation and opened plate pack for (a) parallel arrangement (1 × 2/1 × 2) and (b) series arrangement (2 × 1/2 × 1)

2. Scraped-Surface Heat Exchangers

Scraped-surface heat exchangers (SSHEs) are widely used in industries that manufacture and thermally process fluids. In particular, the food industry makes great use of such devices. One of the food industry's constant goals is to reduce manufacturing costs whilst retaining control over food quality. The process optimisation of heat transfer is therefore a particular priority for the many food products that are made by heating or cooling a raw material or mixture of materials.

A number of distinct types of heat exchangers are routinely used in the food sector. The fluid flow and heat transfer in classical plate and tubular heat exchangers is normally relatively straightforward and easy to understand and optimise. An illustration of SSHE is shown in Figure 19.8.

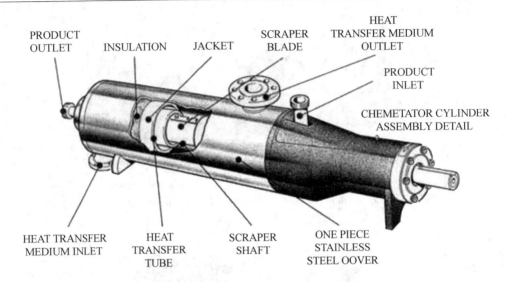

Figure 19.8 Diagram of a typical SSHE

3. Shell-and-Double Concentric-Tube Heat Exchangers

The tube heat exchangers is a typical type of heat exchanger, which can be used in many specific applications such as air conditioning, waste heat recovery, chemical processing, pharmaceutical industries, power production, transport, distillation, food processing, cryogenics, etc..

There are shell-and-tube heat exchangers and simple or counter cross-flow heat exchangers for several decades. The simple or grooved tubes can have fins with varied forms. These heat exchangers can work in simple or in two-phase (condensation and evaporation). The heat transfer surface is increased by heat exchanger length unity using double tubes instead of simple tubes. The stream flows through the gap channel between the inner and outer tubes exchanges the heat with two fluids, shown in Figure 19.9.

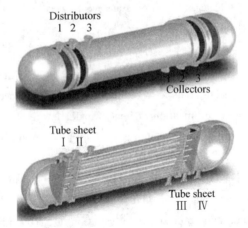

Figure 19.9 Perspective view and longitudinal section of the shell-and-double concentric-tube

4. A Shell and Tube Heat Exchanger

Comparison between smooth and helically corrugated wall tubes, fluid foods are often subjected to thermal treatment inside surface heat exchangers. Besides the need for high heat transfer performance, also low friction losses and easy cleaning and sanitizing properties of the surface are imperative. In food process industry these requirements are often met by the shell and tube heat exchanger equipped with helically corrugated walls(Figure 19.10).

The work concerns convective heat transfer and friction losses in helically enhanced tubes for both Newtonian and non-Newtonian fluids, such as several of fluid foods, whole milk, cloudy orange juice, apricot and apple puree in a shell and tube heat exchanger. Considering both fluid heating and cooling conditions, the helically corrugated tubes are particularly effective in enhancing convective heat transfer for generalized Reynolds number ranging from about 800 to the limit of the transitional flow regime.

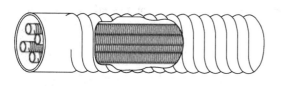

(a) Scheme of a tube bundle   (b) Helically corrugated tube

Figure 19.10 The shell and tube heat excharger equipped with helically corrugated walls

19.6 Extrusion Technologies

Extrusion technologies have an important role in the food industry as efficient manufacturing processes. Their main role was developed for conveying and shaping fluid forms of processed raw materials, such as doughs and pastes. Extrusion cooking technologies are used for cereal and protein processing in the food and, closely related, pet foods and feeds sectors. The processing units have evolved from simple conveying devices to become very sophisticated in the past decades. Today, their processing functions may include conveying, mixing, shearing, separation, heating or cooling, shaping, co-extrusion, venting volatiles and moisture, flavour generation, encapsulation and sterilisation. They can be used for processing at relatively low temperatures, as with pasta and half-product pellet doughs, or at very high ones with flatbreads and extruded snacks. The pressures used in extruders to control shaping, to keep water in a superheated liquid state and to increase shearing forces in certain screw types, may vary from around 15 to over 200 atmospheres.

19.6.1 Extrusion Process

The most important feature of an extrusion process is its continuous nature. It operates in a dynamic steady state equilibrium, where the input variables are balanced with the outputs. Therefore, in order to obtain the required characteristics in an extrudate, the multivariate inputs must be set at the correct levels to give the dependent physical conditions and chemical process changes within the barrel of the machine.

This new understanding within the industry has led to better use of existing machines and modification to improve their function. The development of extruders has moved forward from a purely empirical approach, which led to the development of products on single screw and twin screw extruder machines. Extrusion technologists are now more likely to use mathematical modelling on different applications of extruders.

Today, extruders come in a wide variety of sizes, shapes and method of operation. There are three types of food extruder found in industry: hydraulic ram, roller, and screw type extruders. The screw extruders are very deferent to the other two having special features such as continuous processing and mixing ability. Single and twin screw types are both widely used in the food process industry. Many raw materials are available to be processed into nutritious and palatable food items.

19.6.2 Design Approach

In the design process, the level of pressure likely to be developed within the extruder barrel becomes the most important factor as it determines the sizes of all the important components of the extruder such as bearings, shaft diameters and barrel dimensions. As these machines often operate under elevated temperatures, pressure tends to vary unpredictably because of complex rheological properties of food dough as related to varying temperature along the barrel. Therefore, extruder process variables (such as die sizes, pressures, temperatures etc.) were made on the chosen range of food materials. The central parts of an extruder is the screw, barrel and dies.

19.6.3 Single and Twin Screws

Single screw extruders (Figure 19.11) have simpler construction and are generally cheaper. Single screw machines meet most of the criteria for simple requirements. But they are more likely to block than twin screw extruders. Particularly in the case that adequate flow are fed into the screws and the material flows down the barrel without spinning within it. The forward motion of the food dough relies entirely on the friction between the material and the interior surfaces of both screw and barrel. This is particularly so in a single screw but less so in a twin screw.

In twin screw extruders (Figure 19.12), the screws can be made as either co-rotating or counter rotating, with differing amounts of intermesh, pitch and clearances, all of which affect the processing characteristics of the machine. Since the flight of one screw engages the channel of the other the twin screws prevent material from sticking to the screws and rotating with the

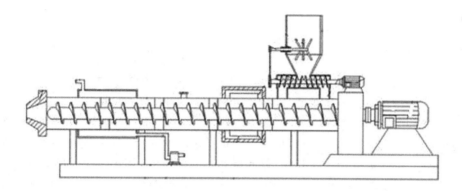

Figure 19.11 Schematic of single screw extruder

screw, hence encouraging forward motion of dough with reduced slip. This also enhances the mixing of materials in the channels of the screws. The counter rotating type of twin screw extruder can give better material flow characteristics pushing the material positively forward towards the die but reducing the degree of mixing of materials within the extruder. The dilemma facing the designer is that although the operational problems can be avoided by using twin screws their sophisticated constructional features would have to be simplified significantly to achieve the low cost objective.

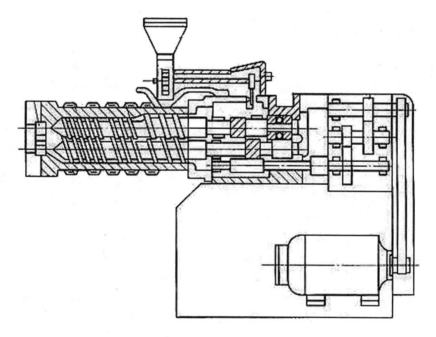

Figure 19.12 Schematic of twin screw extruder

19.7 Biscuit-Making Machine

In many areas, wheaten food is consumed for which the dough is developed by a sheeting roll process, such as bread, cookies, flour tortillas and Chinese noodles.

All mechatronics are automatically completed, from feeding, forming, drying, conveying, fuel injection, cooling, waste recycling and so on. The food materials process such as doughs is a broad area of food industrial activity. Due to the difference in the visco-elastic properties of solid foods, and doughs in particular, the materials actually used show complex time-dependent responses during the forming processes. The following are typical components of biscuit-making machine.

Although flour particles are sufficiently hydrated after mixing and resting, the development of the gluten matrix is far from complete and is localized without continuity. It is during the sheeting process that the continuous gluten matrix is developed. Under compression, adjacent endosperm particles become fused together so that the protein matrix within one endosperm particle becomes continuous with that of adjacent particles. The sheeting process is intended to achieve a smooth dough sheet with desired thickness, and a continuous and uniform gluten matrix in the dough sheet.

19.7.1 Sheeting Machines

The sheeting of doughs using roll mills is a prominent operation in the production of both biscuits and bread. In many operations, rolling is well understood, for instance, metal sheeting is considerable; and the rolling of polymer melts, called calendaring, has also been studied. From the mechanical point of view, sheeters, gauge rolls, and laminators are all devices for reducing the dough thickness by compressing and gauging the dough mass into a sheet by means of chain rolls. They are used in the biscuit production lines, and they vary. Here is a general instruction and some parameters in sheeting process.

19.7.2 Contact Angle

In the sheeting process, the deformation takes place as the dough comes into contact with the rolls. Figure 19.13 shows the deformation area where D is the sheeting roll diameter, $D = 2R$; h_0 is the thickness of the sheet before sheeting; h_1 is the sheet thickness after sheeting; Δh is the absolute reduction of the sheet thickness, $\Delta h = h_0 - h_1$; L_0 is a part length before sheeting; L_1 is the length of the part after sheeting; b_0 is the width of a part before sheeting; b_1 is the width of the part after sheeting; Δb is the absolute width increment of the part, $\Delta b = b_1 - b_0$; α is contact angle; and e is the gap between two rolls, theoretically, $e = h_1$.

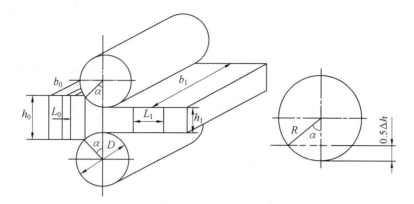

Figure 19.13 The deformation area in sheeting process

19.7.3 Cutters

The early type of reciprocating cutter worked in a vertically reciprocating mode by which the dough sheet was cut into pieces as it was moved forward step-by-step by a set of pawl and ratchet wheel mechanisms. This type of cutting machine usually works at a low speed, with a low machine output. The rotary cutter was invented in the 1970s. It completely changed the biscuit cutting method from the traditional reciprocating action into a rotary mode. With the development, the third generation of rotary cutter is the four-roll rotary cutter unit which comprises a pair of driven rolls, consisting of a printing and a cutting roll with their respective rubber covered anvil rolls. In the production line, a typical rotary cutting machine has with two top rolls and cutting web with one or two anvil rolls.

19.7.4 Rotary Moulders

Biscuit machine can produce various kinds of cookies. Cookies, as the term is used here, means the kind of baked cereal food shaped in one operation without sheeting before shaping. The doughs from which cookies are produced are not suitable for sheeting. According to their formulation and consistency, the doughs can be classified into short and soft doughs. The cookies are classified into short dough cookies and soft dough cookies. Rotary moulders are suitable for short doughs, and extruders, including depositors, are suitable for soft doughs. The rotary moulding machine, or rotary moulder for short can be used in the sheeting and cutting process of the extensible hard doughs for crackers and semi-sweet biscuits.

19.7.5 Rotary Moulders

Danish batter cookies, Viennese whirls, and Spritz cookies are rich in fat, brandy snaps are rich in sugar, and Jaffa cakes, sponge boats, and lady's fingers are sponge textured. The doughs of the above-mentioned foods are either relatively viscous, fruited, or have small particle ingredients included or are soft enough to be just pourable. These types of dough are conveniently referred to as soft doughs. The formation of this kind of dough cannot be carried out by sheeting, printing, rotary cutting, or even rotary moulding.

Extruding machines are expressly designed to deal with the formation difficulties of soft dough. "Depositing" is a form of extrusion, so these two methods of dough piece forming are not distinct from one another. A typical extruding machine consists of a dough feed assembly, die assembly, power and transmission, frame, and control system.

19.8 Baking Machine

19.8.1 Introduction

Due to the high consumption of bread, the baking sector constitutes the most important section of the food industry. Frequently changing quality prevents the development of these industries and often leads to consumer dissatisfaction. Consumer demand is one of the most important factors in the production progress and development. Therefore, technological development of the bread industry is done for the purpose of boosting quality. Additionally, the quality of bread and reducing bread waste are extremely important .

Wheat gluten quantity and quality are closely related to bread quality. Addition of high-quality wheat into the blend and the use of permitted baking additives are recommended to obtain a desirable standard of quality. Moreover, a variety of additives are used to increase the nutritional value of the bread and to delay staleness and spoilage. In particular, the additives used in large-scale bread production facilities for the purpose of balancing changes in flour quality significantly affect the dough rheology and bread features. In the baking industry, widely used additives are oxidants, reducing agents, emulsifiers, and enzymes. The roles of components and process steps should be well known in order to correct any faults in the bread and identify the source of the lack or surplus that may arise during production.

Flour quality should always be verified with a baking test in any instance of quality consideration. An automatic bread machine was used to optimize the bread formula and reported promising results. In determining the quality of flour for bread making, a method that is easy to apply and practical is needed. Because of the changes in wheat quality from year to year, the analysis must be repeated and the obtained test results should be compared with each other.

19.8.2 Baking Bread

1. Normal Baking Technology

(1) Dough mixing.

A usual method of converting flour into dough is to mix flour, together with fat and salt, with water, in a small fraction of which pressed bakers' yeast has been suspended. The temperature of the water is adjusted so that the dough after 15 minutes' mixing in a mechanical mixer has a temperature of about 24 ℃ (78 ℉).

(2) Bulk fermentation.

The dough is maintained at its temperature of 24 ℃, covered to prevent the evaporation of moisture, and allowed to ferment for, say, two and a half hours. Varying fermentation times are used in different processes.

(3) Dividing and moulding.

The dough is put through a divider. This is a machine that cuts it into accurately weighed pieces, out of which the individual loaves will be produced. These pieces are rolled mechanically into balls by a machine called a moulder, and dropped by the machine into a series of slowly moving canvas pockets, which travel at a temperature. The gluten fibres, which have been rather roughly stretched by the moulder, recover in this time.

(4) Proving.

The relaxed dough pieces are dropped out of the canvas pockets into a second moulding machine which, in its turn, drops the now sausage-shaped pieces of still-fermenting dough into individual baking tins. These travel slowly through another prover for 40 to 50 minutes, where the temperature is maintained between about 95 ℉ and 100 ℉ (35 ℃ to 38 ℃).

(5) Baking.

If the physics and biochemistry of the whole operation have gone exactly to plan, that is to say, if the fermentative activity of the yeast has been precisely what was expected and the strength of the flour protein has been the same as that for which the times and temperatures of the process used has been worked out, then the dough will have risen to the appropriate height in the tins—and no higher.

In a modern bakery, the prover is so arranged that the emerging tins can be transferred directly into the oven. If a travelling prover is used, then it will be employed in association with a travelling oven. This is commonly adjusted so that the loaves are exposed to an atmosphere of 480 ℉ to 500 ℉ (250 ℃ to 260 ℃), injected with steam, for period of 40 to 50 minutes.

2. Continuous Dough Making

Continuous dough making is a modern development in food technology. A powerful mixer pulls out the protein fibres of the gluten in a single rapid operation in place of the slow stretching brought about by the production of gas bubbles in yeast fermentation.

The dry flour is metered through a pipe into a pre-mixing tank and at the same time a measured amount of a brew of fermenting sugar is pumped in. This brew is prepared about two hours beforehand and is composed of groups of ingredients.

When the brew and the flour have been put together with the right amount of water for the particular flour in use, they are mixed by means of a powerful arrangement of blades to obtain a satisfactory "development" of the gluten.

19.9　Conclusion

Food processing is the process of artificially treating raw foods or other raw materials to produce better or more beneficial foods. With the development of the national economy and the improvement of people's living standards, the people have put forward higher requirements for the food industry. Modern food has developed in the direction of nutrition, green, convenience and functional food. The food industry has also become a pillar industry of the national economy, and the food machinery industry will develop more rapidly.

The development of industrial technology now has a huge impact on food processing. The rapid development of food processing technology has ushered in the prosperity of the industry. Food processing is transforming and upgrading from traditional manufacturing to automation, informatization, and networking. The level, variety and production capacity of food machinery directly reflect the comprehensive strength of the national food industry equipment, which requires domestic food machinery products and food machinery manufacturing technology to reflect informatization, digitization, networking, intelligence, refinement and high speed Change. Design innovation is a promoter of transformation from "Made in China" to "Created in China".

Words/Phrases and Expressions

simplify 简化
baking 烘烤
bread making 制面包
dairy product manufacture 乳制品制造
fortify 加强, 加固, 增强
drying 干燥
thermal processing 热处理
fermentation 发酵
freezing 冷冻
infusion 浸泡
juicing 榨汁
enucleate 去核
malting 麦芽
milling 制粉
parboiling 煮沸
peeling 去皮
peeling and cooking 去皮和烹饪
storage 储存
storage and milling 储存和研磨

washing 洗涤
cuticular 表皮的
detergent 洗涤剂
winemaking 酿酒
horizontal mixer 卧式搅拌机
dehydration 脱水
convection drying 对流干燥
drum drying 转鼓干燥
spray drying 喷雾干燥
fluidized-bed drying 流化床干燥
freeze drying 冷冻干燥
explosive puffing 爆炸性膨胀
osmotic drying 渗透干燥
solute-infused 溶质注入
dielectric drying 介电干燥
hygienic 卫生的
centrifugal pump 离心泵
emulsion 乳浊液, 乳剂
positive displacement pump 容积泵
power pump 动力泵
trite 陈词滥调
High Pressure Processing 高压处理
pulsed electrical field 脉冲电场
decontamination 净化, 去污
high power ultrasound 高功率超声
oscillating magnetic field 振荡磁场
organoleptic 感官的
pasteurise 巴氏杀菌
vessel 容器
autofrettage 自动强化
residual compression stress 残余压缩应力
antibacterial factor 抗菌因子
spore 孢子
pathogen 病原体
heat transfer 传热
convection 对流
conduction 传导
radiation 辐射

heat exchanger 换热器
Tube Heat Exchanger(THE) 列管式换热器
Plate Heat Exchanger (PHE) 板式换热器
Scraped-Surface Heat Exchanger (SSHE) 刮板式换热器
Shelland Tube Heat Exchanger 管壳式换热器
Helically Corrugated Tube 螺旋波纹管
extrusion technology 挤出技术
venting volatiles 排出挥发物
encapsulation 封装
sterilisation 灭菌
single screw extruder 单螺杆挤出机
twin screw extruder 双螺杆挤出机
gluten 面筋
endosperm 胚乳
contact angle 接触角
reciprocating 往复式
pawl 棘爪
ratchet wheel 棘轮
additive 添加剂
oxidant 氧化剂
reducing agent 还原剂
emulsifier 乳化剂
enzyme 酶
informatization 信息化
digitization 数字化
networking 网络化
intelligence 智能化
refinement 精细化

Chapter 20 Chinese Snack Processing

Chinese food occupies a bright and colourful place in the blossomy flower garden of world food. It is a part of China's ancient civilization and long history. In recent years, it has attracted more attention, and is becoming more and more popular. To taste Chinese food is a pleasure of life for the visitors to China. In this part, some typical Chinese snacks and cooking process will be introduced. As you taste the delicious food, you will get much pleasure from it.

20.1 Noodles Processing

Noodles in various contents, formulations, and shape shave been the staple foods for many countries since ancient time. They can be made from wheat, rice, buckwheat, and starches derived from potato, sweet potato, and pulses. Noodles based on wheat are prepared mainly from three basic ingredients: flour, water, and salt. There exist two distinct types of wheat flour noodles based on the presence and absence of alkaline salts, regular salted noodles, and alkaline noodles.

20.1.1 Introduction

Most noodles today are produced by machine. While the actual process for manufacturing a particular type of noodle may different, the basic principles involved are practically the same. Many of these principles stem from the ones used in the making of ancient hand-made noodles. Noodles machines are highly valued for their superior eating quality, presumably due to the mode of gluten formation. Because developments in noodle processing technology, such as vacuum mixing, waved rollers, and multi-layer sheeting, were based on the principles of gluten development in hand-made noodles.

Noodle products are usually made from common wheat fine flour by a process of sheeting and cutting as opposed to pasta products, which are processed from coarse semolina milled from durum wheat by extrusion. The basic process of dough mixing, sheet forming, compounding, sheeting/reduction, and cutting are essentially constant for all machine-made noodles. Noodle strands coming out of cutting rolls can be further processed to produce different types of noodles.

20.1.2 Noodle Unit Basic Processing—From Flour to Raw Noodle Strands

Despite the large variation of formulation, size, and shape of noodles, the process to form noodle strands is remarkably constant for different types of noodles. It typically comprises dough mixing, formation of dough sheets, compounding of two dough sheets, and sheet thickness reduction by rolling, and noodle strand formation by passing the dough sheet through a pair of cutting rolls. After cutting, there is great flexibility in further processing and packaging. The noodle strands could be packed directly and marketed as fresh noodles, or could be dried, steamed, fried, boiled, frozen or undergo a combination of these processes to make different kinds of noodles.

1. Mixing

Mixing is the first step in noodle processing. Most ingredients are pre-dissolved in water and stored in a tank. Wheat flour is weighed and placed into a mixer and the correct amount of mixing water is added. In noodle manufacturing, the main aims of mixing are to distribute the ingredients uniformly and to hydrate the flour particles. There is little gluten development during the mixing stage in the low water absorption noodle dough. The degree of gluten development, however, could be very significant in high water absorption dough (>26%) with long mixing time (>15 min). A properly mixed noodle dough should have its gluten proteins hydrated as much as possible but not to the degree that the dough sheet would be problematic during sheeting due to stickiness. This would maximize the formation of a continuous gluten matrix with embedded starch granules during sheeting.

There are two types of mixers commonly used in the noodle industry: the horizontal mixer and the vertical mixer. They can both provide good mixing and some kneading actions during mixing. Both mixers are usually operated at medium speed (70-100 r/min) for 10-20 min of mixing. There have been a few new mixers developed for the noodle industry: the continuous high-speed mixer; the low-speed super mixer; and the vacuum mixer.

2. Sheeting

Although flour particles are sufficiently hydrated after mixing and resting, the development of the gluten matrix is far from complete and is localized without continuity. It is during the sheeting process that the continuous gluten matrix is developed. Under compression, adjacent endosperm particles become fused together so that the protein matrix within one endosperm particle becomes continuous with that of adjacent particles. The sheeting process is intended to achieve a smooth dough sheet with desired thickness, and a continuous and uniform gluten matrix in the dough sheet.

The dough crumbs are transferred to a hopper and passed through one or two pairs of sheet rolls to form continuous dough sheets. The sheet is reduced in thickness in steps by passing the sheet through a series of sheeting rolls which have a gradually reduced gap between the rolls (Figure 20.1). The speed of each pair of sheeting rolls is controlled based on the linear velocity of the last pair of rolls which is usually limited within 28 m/min.

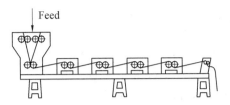

Figure 20.1 Schematic of multiple sheeting

3. Cutting

The width and shape of the noodle strands are determined by the cutting rolls. The cutting device consists of a pair of slotted rolls with identical slot widths. The slots on each roll are offset from one other to allow cutting to occur. The two cutting rolls are aligned horizontally, with the rear one turning clockwise and the front one counter-clockwise at the same speed. Cutting force is generated between the neighboring two sharp edges of the slots of the two cutting rolls. There is a comb underneath each cutting roll to prevent the noodle strands from sticking to the rolls. The shape of the cross-section of the noodle strands depends on the groove of the slot, the width of the slot and the thickness of the dough sheet. Noodle strands are finally cut into proper lengths by a length-cutter. In the case of instant noodle production, the noodle strands are continually fed into a traveling net conveyor which moves slower than the cutting rolls above it. The noodle hob is shown in Figure 20.2.

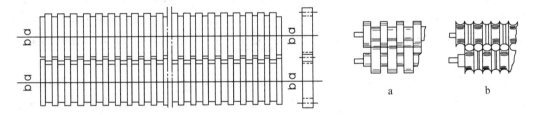

Figure 20.2 Schematic of noodle hob ((a) Square hob, (b) round hob)

20.1.3 Noodle Unit Secondary Processing—From Raw Noodle Strands to Finished Products

1. Drying

The shelf life of noodles can be significantly extended if microbiological and biochemical stability is ensured. The most effective way of achieving this goal is to dry the noodle to a moisture content at which microbiological growth is impossible. Noodle moisture can be removed by air-drying, deep-frying, or vacuum-drying.

Noodle quality has to be preserved during the drying process. Improper drying could damage the noodle structure, causing over-elongation, cracking, warping and splitting of noodle strands. These conditions result in problems in handling and packaging. Moreover, cooking properties and texture could be severely affected. A three-stage drying process, involving predrying, drying, and cooling, is a very common practice.

2. Steaming

Steaming is widely used in noodle processing. As long as the temperature of the steam is high enough, starch gelatinization and protein denaturation occur in wet raw noodles during steaming. The degree of cooking depends on the original moisture content of the noodle; the amount, pressure and temperature of the steam; and steaming time.

In order to produce high quality steamed noodles, it is very important to have raw noodles made from dough with high water absorption, and to use saturated wet high temperature steam in the steaming process. Under-steamed noodles will have a hard core and will be difficult to cook properly by stir-frying before serving. Over-steamed noodles are soft and sticky. The desirable moisture content of steamed noodles for stir-frying is 59%–61%.

3. Boiling

Boiling is a simple process but very critical in terms of finished product quality. The application of boiling in noodle processing has increased significantly in recent years due to the increased popularity of chilled, frozen, and long-life noodles. Boiling time depends on the size of the noodle strands and the types of finished products. A proper moisture content and moisture gradient in the noodle strand is the key to the texture of finished product quality. High water absorption and high salt content (up to 8%) in noodle processing can shorten the necessary boiling time, and therefore, decrease the cooking loss. Noodles made from flour with low starch damage have lower cooking losses than those made from flour with high starch damage. While hardness of water has no significant effect on cooking loss, the alkalinity of the water does.

4. Freezing

The texture of boiled noodles deteriorates very fast due to the disappearance of moisture gradient between the interior and exterior of noodle strands during storage. Fast freezing, however, can extend the "tastiest" state of boiled noodles. After boiling, noodles are first washed with cool water, then immersed in cold water under 5 ℃, and finally fast frozen by blasting cold air of −30 ℃. Noodle strands are easier to separate during thawing if they are cooled to 0–5 ℃ before fast freezing. Over-freezing (< −40 ℃) could damage the noodle structure because the expansion of the noodle core during freezing can break the noodle surface, which freezes completely before the noodle core is frozen.

20.2 Steamed Bun Manufacture

In China, even in Southeast Asia, the "steamed bun" is as important as baked bread in the West. The bun is also referred to as "steamed bread", which is also a fermented food. But its technology is not as time-consuming and complicated as the baked bread process. The steamed bun is called "Man Tou" or "Mo" in Chinese. It can be used for sandwiches, toast, and other types of food. The bun can be described as soft bun filled with sweet or savory ingredients. Typically produced from leavened dough, the steamed bun can be made as a seasoned bun to hold a filling of cooked meat and vegetables blended into a semi-smooth texture or it may be made as a sweet bun containing a filling of pureed fruits or sweet bean pastes. There are also filled buns, for example red bean, chicken curry and meat curry fillings. The "Xi'an Pao Mo" is very famous, and is popular with the Chinese and with many foreign visitors. "Xi'an" is the name of Shannxi province capital which was the capital of the Tang Dynasty in a period of great prosperity in China. "Pao" means to soak the pieces of bun (Mo) in a very tasty and special soup.

The dough (with the ratio of flour to water 2 : 1) is traditionally fermented by means of a small piece of leavening dough. Nowadays, some yeast is employed to shorten the fermentation time from some hours to about forty minutes. When made in this way, the finished buns have a slightly different flavour from the traditional product, but they are very soft and tasty.

Doughs fermented in the traditional way must first be neutralized by sodium carbonate or sodium bicarbonate, and then left to stand for a while before forming into buns either by hand in the domestic kitchen or by special machinery in the restaurant kitchen. The well-rounded dough pieces will be put into a steam cooker. Twenty minutes later, the dough balls are well-steamed into white, smooth, appetizing buns with a volume two to three times that of the original dough piece.

Steaming is one of the methods in transferring heat in food cooking process. The term is also defined as cooking of prepared foods by steam (moist heat) under varying degrees of pressure. Steam is the vapour forms when water is heated to its boiling point. Based on the Figure 20.3, several phases are completed before water becomes steam; starting from the first condition from ice where heat is prepared until 32 ℉, as ice will be turned into water.

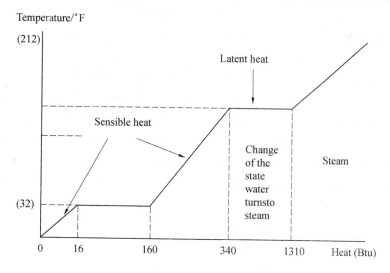

Figure 20.3 The process of steam creation

With a continuous heating, the water will become latent with the heat and eventually turns into steam after the heat approach certain level. Once the steam was produced, the steam will be trap in the steam trap which removes air and condensate from steam lines and heating units. Steaming method has few benefits, where the method able to preserve the nutrition value of the food as well as to make the food lighter and easier to digest. These factors make steaming method becoming more relevant and important to today's health-conscious consumers.

20.3 Dumpling Machine

A "dumpling" is called "Jiao Zi" in Chinese. It is a favourite food in China. Traditionally, "Dumpling" is a symbol of reunion, happiness, and safety. When one of your relatives or friends is going on a journey, you may make dumplings as a send-off dinner, which expresses your wish that the traveller will arrive "safe and sound". The Chinese New Year is the most important festival in China, just like Christmas in many Western Countries. This is a day of family reunion, at its midnight advent the whole families always sit together, around a big table, and have their New Year dumplings after setting off fireworks and firecrackers. Here, the "dumpling" symbolizes reunion, harmony, and happiness. As well as the specialized meaning at a special time, people often make dumplings when receiving family guests or for their own enjoyment, since it is such a delicious food.

The manual procedure is time-consuming. Along with the accomplishment of dumpling-making mechanization, people can now buy frozen dumplings in the supermarkets. Dumplings more and more often appear on the domestic dining tables. Many restaurants make dumplings as a snack (quick meal) which is becoming more and more popular.

A Chinese dumpling comprises two parts: one is the casing dough, the other is the filling, a complex mixture of minced meat or seafood, vegetables, salt, edible oil, ginger, pepper powder, and other flavourings. The dough is a simple mixture of flour and water at a ratio of 100 (flour) : 38-40 (water).

For hand-made dumplings, the casing material (dough) should be rolled into small round sheets with a weight of about ten grams. For example, 100 g of flour is mixed with 40 g of water, the dough will be 140 g which can be divided into 12 to 18 pieces.

For a typical dumpling production line, a vegetable-washer, processor (cutter), dehydrator, and two mixers, of which one is for dough-mixing and the other for stuffing-mixing, are needed. A higher productivity can be achieved. The overall principle structure is shown in Figure 20.4, and the produced dumplings are in Figure 20.5.

(1) The dough is continuously pressed by several rollers for several times, and cut into a dough belt with a thickness of about 15 mm and a width of 85 mm.

(2) The thick dough belt is introduced into a 2-stage pressing rollers into thin dough belt with the thickness of dumpling skin.

(3) The thin dumpling skin combined into a dough tube with a burr by a pair of kneading

rollers in the forming part.

(4) The dough tube is kneaded into a completely closed form by a pair of kneading and cutting knives. At the same time, the excess dough is cut off and is led out by the guide roller and recovered.

(5) The filling pump continuously and evenly feeds the filling into the tubular dough.

(6) The stuffed tubular dough is cut and wrapped into dumplings by the in opposite rotating and cutting of the forming roller and auxiliary roller.

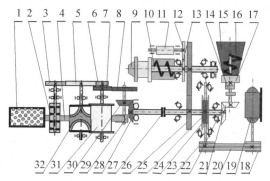

1-slide plate; 2-powder duster; 3-gear; 4-vibrating rod; 5,6,7,8-gears; 9-feeding mouth; 10-feeding auger; 11-auxiliary pressure roller; 12,13-gears; 14-clutch; 15-stuffing hopper; 16-stuffed auger; 17-vane pump; 18,21,22,23-belt wheels; 19-motor; 20-belt; 24-worm; 25-worm wheel; 26-gear; 27-coupling; 28,30 gear; 29-cam;31-bottom roller; 32-forming die roller

Figure 20.4　The overall structure

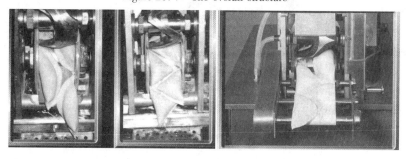

Figure 20.5　The result of production

The main components are shown as follows.

1. Fill stuffing mechanism

A volumetric vane pump is used to complete the filling operation (Figure 20.6).

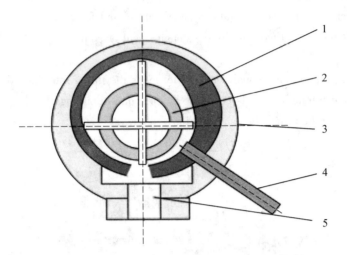

1-Stator; 2-Rotor sleeve; 3-Blade; 4-Adjusting handle; 5-Stuffing outlet
Figure 20.6 Fill stuffing mechanism

2. Conveying dough and making wrappers mechanism

The continuous extrusion of the tubular dough meeting the requirements is completed through the face auger and the inner and outer nozzles (Figure 20.7).

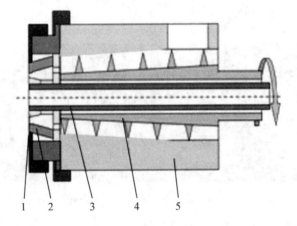

1-Inner mouth; 2-Outer mouth; 3-Stuffing tube; 4-Inverted auger; 5-Barrel
Figure 20.7 Dough conveying mechanism

3. Roll cutting forming mechanism

Stuffed tubular dough is made by an enema forming mechanism, and then it is continuously processed into a dumpling green body by a roll cutting method (Figure 20.8).

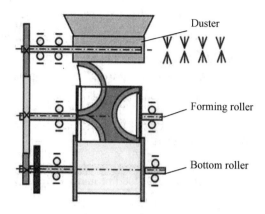

Figure 20.8 Roll cutting mechanism

20.4 Hun Tun Processing

Hun Tun is another stuffed snack, very popular in China. The differences between dumplings and Hun Tun are not only in the shape but also in the method of cooking, the hardness of the casing dough, and the consistency of the stuffing materials. Dumplings are cooked in boiling water or fried in a pan, but Hun Tun are usually cooked in a soup which may be chicken soup, or seafood soup, or some other tasty clear soup, dependent on personal taste. Dumplings are often eaten with a sauce (a mixture of vinegar and other flavourings such as mustard), while Hun Tun are already flavoured by the soup in which they have been cooked. A bowl of Hun Tun is always a good choice for breakfast or a midnight snack.

The casing material of Hun Tun is also a mixture of flour and water with a ratio of 100 (flour) : 35 or less (water), while for dumpling dough it is about 100 : 38-40, and for steamed buns, the dough is about 100 : 50. So the Hun Tun dough is the hardest, and no dusting flour is needed in Hun Tun production.

Usually, the stuffing for Hun Tun is a mixture of meat of seafood with various flavourings, but seldom vegetable, so that it is rather sticky.

The formation of Hun Tun is done in the following stages:

(1) prepare casing and filling materials-dough and stuffing;

(2) roll the dough into a sheet with a thickness less than 2 mm;

(3) cut the dough sheet into trapezium pieces with dimensions of about 60 mm×60 mm× 110 mm (upper×height×bottom);

(4) fill the piece of dough sheet with stuffing;

(5) fold and close it.

The first Hun Tun-making machine, Hun Tun maker for short, was invented in about 1980. After several years of work by many technicians and engineers, the newly developed Hun

Tun maker is an advanced and ingeniously constructed machine in which the shaping process is an imitation of the manual method. It is about the size of a domestic microwave oven, but is very efficient. It is capable of forming 4 920 pieces of Hun Tun per hour. The weight of the products can be changed easily within a range of 7 g to 15 g. This space-saving and energy-saving machine can be used in restaurants and food factories.

20.5 Sweet Dumpling Processing

Sweet dumplings are the favorite traditional food of many Chinese people, as their meaning is reunion in China. Whether celebrating festivals or entertaining guests, sweet dumplings are preferred.

Especially in the Lantern Festival, every household would boil sweet dumplings to celebrate the holiday. Many sweet dumplings are made from water and glutinous rice flour containing largely amylopectin (98% of total starch) and little amylose (less than 2% of total starch). For thousands of years, wet milling of the glutinous rice following an overnight soak in water is commonly used as a material processing stage in traditional process. However, the long soaking time results in bacterial and harmful microorganisms' growth in rice, leading to nutrient losses and uneven product quality, thus restricting the large-scale industrial production of sweet dumplings.

During the milling of early polished rice, dry milling method was used to break the rice, but dry milling generated more thermal and mechanical energy, which led to an increase in the damage to starch granules, so dry-milled rice flour exhibited a higher degree of starch damage in comparison to wet-milled rice flour. Although dry milling presents fewer health risks and improves the stability of rice products, and does not produce waste water, the eating quality of sweet dumplings produced from dry-milled glutinous rice flour is not acceptable to consumers. Because rice proteins do not form a stable network structure, the viscoelastic quality of sweet dumplings depends primarily on the properties of the starch component. The addition of strengthening agents is a convenient and inexpensive method for improving the quality of dry-milled glutinous rice flour, but this may result in the overuse of additives. So it is necessary to improve the adverse effects of dry milling on the characteristics of glutinous rice flour and cooking qualities of sweet dumplings.

In sweet dumplings machine, the main component is the roller plate, which is tilted at an angle of about 15 degrees. The big and round plate continuously rotates. The stuffing balls roll on the roller plate and are covered with flour to form round sweet dumplings.

For cooking qualities of sweet dumplings, the paste soup dumpling soup rate is expressed by turbidity. The higher the light transmission is, the more transparent the soup is, and the less the sediment is, the better the quality is. Cooking loss was calculated by the following formula:

Cooking loss = Dry weight of the cooking water/Weight of uncooked rice dumplings

(20.1)

Different types or varieties of rice have different moisture contents. It is reported that semi-dry milling at 30% moisture may provide the protective effects on the characteristics of rice flour and qualities of rice noodles similar to wet milling.

Words/Phrases and Expressions

snack 小吃
buckwheat 荞麦
slotted roll 开槽辊
elongation 伸长
crack 破裂
warp 翘曲
split 分裂
steaming 蒸
starch gelatinization 淀粉糊化
protein denaturation 蛋白质变性
core 内芯
alkalinity 碱度
steamed bun 馒头
neutralize 中和
sodium carbonate 碳酸钠
sodium bicarbonate 碳酸氢钠
glutinous rice 糯米
starch 淀粉
amylopectin 支链淀粉
amylose 直链淀粉
viscoelastic 黏弹性的

Chapter 21 The Use of Ultrasound and Microwaves

21.1 Ultrasound in Food Processing

Ultrasound is well known to have a significant effect on the rate of various processes in the food industry. Using ultrasound, full reproducible food processes can now be completed in seconds or minutes with high reproducibility, reducing the processing cost, simplifying manipulation and work-up, giving higher purity of the final product, eliminating post-treatment of waste water and consuming only a fraction of the time and energy normally needed for conventional processes. Several processes such as freezing, cutting, drying, tempering, bleaching, sterilization, and extraction have been appliedefficiently in the food industry. The advantages of using ultrasound for food processing, includes: more effective mixing and micro-mixing, faster energy and mass transfer, reduced thermal and concentration gradients, reduced temperature, selective extraction, reduced equipment size, faster response to process extraction control, faster start-up, increased production, and elimination of process steps. Food processes performed under the action of ultrasound are believed to be affected in part by cavitation phenomena and mass transfer enhancement. The following presents the knowledge on application of ultrasound in food technology including processing, preservation and extraction. It provides the necessary theoretical background and some details about ultrasound the technology, the technique, and safety precautions. We will also discuss some of the factors which make the combination of food processing and ultrasound in the field of modern food engineering.

Nowadays, the principle aims of modern processed foods technologies, including manufacturing, processing and packaging are to reduce the processing time, save energy and improve the shelf life and quality of food products. Thermal technologies (radio frequency and microwave heating), vacuum cooling technology, high pressure processing and pulsed electric field technology are those novel technologies who have potential for producing high-quality and safe food products but current limitations related with high investment costs, full control of variables associated with the process operation, lack of regulatory approval and importantly consumer acceptance have been delaying a wider implementation of these technologies at the industrial scale.

Utilization of ultrasound in food technology for processing, preservation and extraction is such a system that has evolved to keep the wheel of development rolling. Ultrasound offers a net advantage in term of productivity, yield and selectivity, with better processing time, enhanced quality, reduced chemical and physical hazards, and is environmentally friendly. Food products, such as fruit and vegetables, fat and oils, sugar, dairy, meat, coffee and cocoa, meal and flours, are complex mixtures of vitamins, sugars, proteins and lipids, fibres, aromas, pigments, antioxidants, and other organic and mineral compounds. Nowadays, its emergence as green novel technology has also attracted the attention to its role in the environment sustainability. Ultrasound applications are based on three different methods:

(1) Direct application to the product;
(2) Coupling with the device;
(3) Submergence in an ultrasonic bath.

There are a large number of potential applications of high intensity ultrasound in food processing of which a number are discussed below (Table 21.1).

Table 21.1 Applications of ultrasound in food processing

Applications	Conventional methods	Ultrasound principle	Advantages	Products
Cooking	Stove Fryer Water bath...	Uniform heat transfer	Less time Improving heat transfer and organoleptic quality	Meat Vegetables
Freezing/ crystallization	Freezer Freezing by immersion, by contact...	Uniform heat transfer	Less time Small crystals Improving diffusion Rapid temperature decreasing	Meat Vegetables Fruits Milk products
Drying	Atomisation Hot gas stream Freezing Pulverisation	Uniform heat transfer	Less time Improving organoleptic quality Improving heat transfer	Dehydrated products (fruits, vegetables...)
Pickling/ marinating	Brine	Increasing mass transfer	Less time Improving organoleptic quality Product stability	Vegetables Meat Fish Cheese
Degassing	Mechanical treatment	Compression -rarefaction phenomenon	Less time Improving hygiene	Chocolate Fermented products (beer...)
Filtration	Filters (membranes semi-permeable...)	Vibrations	Less time Improving filtration	Liquids (juices...)
Demoulding	Greasing moulds Teflon moulds Silicon moulds	Vibrations	Less time Reducing products losses	Cooked products (cake...)

Table 21.1 (Continued)

Applications	Conventional methods	Ultrasound principle	Advantages	Products
Defoaming	Thermal treatment Chemical treatment Electrical treatment Mechanical treatment	Cavitation phenomenon	Less time Improving hygiene	Carbonated drinks Fermented products (beer...)
Emulsification	Mechanical treatment	Cavitation phenomenon	Less time Emulsion stability	Emulsions (ketchup, mayonnaise...)
Oxidation	Contact with air	Cavitation phenomenon	Less time	Alcohols (wine, whisky...
Cutting	Knives	Cavitation phenomenon	Less time Reducing products losses Accurate and repetitive cutting	Fragile products (cake, cheese...)

21.1.1 Cooking

In a conventional cooking method, when foods are exposed to elevated temperatures, the outside may be overcooked with the interior insufficiently cooked, and this will lead to a reduction in the quality of the product. Ultrasound has the ability to provide improved heat transfer characteristics, which is a key requirement to avoid such problems, and these have been utilized in cooking. Ultrasound cooking resulted in greater cooking speed, moisture retention and energy efficiency, suggesting that it could provide a new, rapid, energy-efficient method that may improve the textural attributes of cooked meat. Ultrasound cooking provides a significantly faster cooking rate and higher post-cooking moisture content and sensory panellists indicated greater myofibrillar tenderness. This suggests that possible uses of ultrasound in the food processing or food service industries are in the provision of moist precooked or cooked meats for use in prepared meals.

21.1.2 Freezing and Crystallization

Freezing and crystallization are linked in that both processes involve initial nucleation followed by crystallization. Sonication is thought to enhance both the nucleation rate and rate of crystal growth in a saturated or supercooled medium by producing a large number of nucleation sites in the medium throughout the ultrasonic exposure.

Under the influence of ultrasound, conventional cooling provides much more rapid and even seeding, which leads to a much shorter dwell time. In addition, since there are a greater number of seeds, the final size of the ice crystals is smaller and so cell damage is reduced. Accelerated cooling is achieved by improving heat transfer. Acoustic cavitation also occurs and acts as nuclei for crystal growth or by the disruption of nuclei already present. There are clear advantages for the food industry. The technology has been applied to the crystallization of materials such as milk fat, triglyceride oils such as a vegetable oil and ice cream.

21.1.3 Drying

Traditional methods for desiccating or dehydrating food products by a forced stream of hot air are reasonably economical, but the elimination of the interior moisture takes a relatively long time. Moreover, high temperatures can damage the food, which in certain cases may change the colour, the taste and the nutritional value of the rehydrated product. Diffusion at the boundary between a suspended solid and a liquid is substantially accelerated in an ultrasonic field and heat transfer is increased by approximately 30% –60% depending on the intensity of the ultrasound. Ultrasonically assisted air drying continues to be the focus of considerable attention, with new developments now appearing in the design of fluidized bed-type systems. Ultrasonic osmotic dehydration technology uses lower solution temperatures to obtain higher water loss and solute gain rates. Due to the lower temperatures during dehydration and the shorter treatment times, food qualities such as flavour, colour and nutritional value remain unaltered. Ultrasound has also been used as a pretreatment prior to the drying of a range of vegetables. The treatment produced a reduction in subsequent conventional and freeze-drying times and also in rehydration properties.

21.1.4 Brining, Pickling and Marinating

Pickling and marinating are used for a wide variety of vegetables and meat products. Ultrasound allows the pickling time of products to be reduced considerably, particularly those foods with a crunchy texture. It was found that the water and NaCl contents of samples after treatment were higher in sonicated than non-sonicated samples. Moreover, ultrasound reduces the salting time, the formation of a crust and unwanted colouring of raw meat. In the cheese industry, the effect of ultrasound on mass transfer during cheese brining has been investigated. The rate of water removal and sodium chloride gain increased when ultrasound was applied in comparison with brining performed under static or dynamic conditions, suggesting that ultrasound improves both external and internal mass transfer.

21.1.5 Degassing/Deaeration

A liquid contains gases as a mixed condition, such as dissolved oxygen, carbon dioxide, and nitrogen gas. Two common method used for degassing are boiling and reducing pressure while ultrasound has an advantage in the small temperature change. Degassing in an ultrasonic field is a highly visible phenomenon when ultrasound, e.g. an ultrasonic cleaning bath, is used with regular tap-water inside. It occurs when the rapid vibration of gas bubbles brought them together by acoustic waves and bubbles grow to a size sufficiently large to allow them to rise up through the liquid, against gravity, until they reach the surface.

In the food industry, this technique can be used to degas carbonated beverages such as beer (defobbing) before bottling. Compared with mechanical agitation, the ultrasonic method decreases the number of broken bottles and overflow of the beverage. The application of relatively low-intensity ultrasound during the fermentation of sake, beer and wine resulted is a reduction in the time required by 36% - 50%. Ultrasonically assisted degassing is particularly rapid in aqueous systems, but the removal of gas is much more difficult in very viscous liquids such as melted chocolate.

21.1.6 Filtration

In the food industry, the separation of solids from liquids is an important procedure either for the production of solid-free liquid or to produce a solid isolated from its mother liquor. But the deposition of solid materials on the surface of filtration membrane is one of the main problems. The application of ultrasonic energy can increase the flux by breaking the concentration polarisation and cake layer at the membrane surface without affecting the intrinsic permeability of membrane. The liquid jet serves as the basis for cleaning, and some other cavitational mechanisms lead to particle release from the blocked membrane. Ultrasound can also be applied to the production of fruit extracts and drinks. In the case of juice extraction from apple pulp, conventional belt vacuum filtration achieves a reduction in moisture content from an initial value of 85% to 50%, whereas electroacoustic technology achieved 38%. With ultrasound aiding in, there is a significant enhancement in flow rate, preventing blockage of the filter to flow through by lowering the compressibility of both the initial protein deposit and the growing filter cake. Additionally, the combination of filter with ultrasounds increases its filter life, as clogging and caking are prevented by continuous cavitation at the filter's surface. While, there is still need in its development at commercial level.

21.1.7 Demoulding and Extrusion

Generally, the industrial cooking of foods leads to adhesion of the products to the cooking vessel or in other operations it must detach from its mould. To counteract this difficulty in industrial processing of moulded food products, the moulds are fabricated with a surface coating made of a thin layer of silicone or PTFE (polytetrafluoroethylene) or by a coating of white grease. However, it is necessary to replace the mould covering periodically because the shelf-life of the thin lubricating layer is relatively short. Such operations are expensive and, if not 100% successful. At present, to solve this problem mechanical methods such as knocking vibration are used to remove the products. An alternative solution to these conventional methods is to release food products by coupling the mould to a source of ultrasound. The device for demoulding industrial food products couples the mould and the ultrasonic source in order to enhance removal of the product contained in the latter by virtue of the high-frequency relative movement between the contact surfaces of the mould and of the product contained in the latter.

21.1.8 Defoaming

Foam is a dispersion of gas in liquid, with a density approaching that of the gas, i.e., in which the distances between the individual bubbles are very small. Foams find applications in a wide range of industrial processes, including food production and cosmetics. Intensive foaming in the majority of technological processes has the negative consequences expressing in decrease use of useful volume of the technological equipment, infringement of the rules of manufacture and sterility of biotechnological processes, in increase in losses of products and decline of productivity of the equipment, pollution of the environment.

In food manufacturing, foam has historically been controlled by the use of mechanical breakers, lowering packaging container temperatures below the ambient environment, or by the addition of chemical anti-foams. Nevertheless, problems remain, for example: mechanical methods are effective only for coarse foams.

High-intensity ultrasonic waves offer an attractive method of foam breaking since they avoid the need for high air flow, prevent chemical contamination and can be used in a contained environment, i.e., under sterile conditions. This makes it particularly appropriate for implantation in the food and pharmaceutical industries.

A system for ultrasonic defoaming has been developed based on a new type of focused ultrasonic generator. For the perfect defoaming effect, it is not only important to overcome on the acoustic intensity, but also a minimum treatment time is required.

21.1.9 Emulsification/Homogenization

Emulsification is an important mean to deliver the hydrophobic bioactive compounds into a range of food products. Acoustic emulsification offers the following improvements over conventional methods:

(1) The emulsion produced has particles in the sub-micron range with an extremely narrow particle size distribution;

(2) The emulsions are more stable;

(3) Addition of a surfactant to produce and stabilize the emulsion is not necessary;

(4) The energy needed to produce an emulsion by acoustic waves is less than that needed in conventional methods.

21.1.10 Cutting

The introduction of ultrasound in food cutting has improved the performance of overall food processing. Ultrasonic food cutting equipments provide a new way to cut or slice a variety of food products that streamlines production, minimizes product waste and lowers maintenance costs. Ultrasonic cutting uses a knife-type blade attached through a shaft to an ultrasonic source. The ultrasonic cutting characteristics depend on the food type and condition, e. g., frozen or thawed. The most widespread application of ultrasound is in the cutting of fragile foodstuffs. It uses in the particular cases of fragile and heterogeneous products (cakes, pastry and bakery products) and fatty (cheeses) (Figure 21.1) or sticky products. Since the vibration prevents the adherence of the product on the blade and thus reduces the development of micro-organisms on the surface, ultrasonic vibrations provide "auto-cleaning" of the blade.

Figure 21.1 Difference in cutting slices of a bakery product cut with and without ultrasound

In the food industry, Depolymerization, Defrosting/Thawing, Meat Tenderization, Sterilization/Pasteurization, Miscellaneous Effects are also active areas used of ultrasound.

21.2 Ultrasound in Food Preservation

Consumers of the food industry have become more concerned with the hygiene-related aspects of food manufacture. Some say that it is a very art to achieve the best in food preservation. Traditionally heat treatment, in occurrence pasteurization and sterilization, was a method of choice having the ability to destroy both micro-organisms and enzymes, the latter being responsible of food deterioration. However, its effectiveness is dependent on the treatment temperature and time and this leads to loss of nutrients, development of undesirable flavours, colours. As a result, new non-thermal technologies are receiving great interest.

Ultrasound processing is one of these new methods. While its application in food processing is relatively recent, it has been proved that high-intensity ultrasonic waves can rupture cells and denature enzymes, and that even low-intensity ultrasound is able to modify the metabolism of cells. In combination with heat, ultrasonication can accelerate the rate of sterilization of foods, thus lessening both the duration and intensity of thermal treatment and the resultant damage. The advantages of ultrasound over heat sterilization include: the minimizing of flavour loss, greater homogeneity, and significant energy savings.

21.3 Ultrasound-Assisted Extraction

Shortcomings of existing extraction technologies, like increase consumption of energy (more than 70% of total process require energy), high rejection of CO_2 and more consumption of harmful chemicals, have forced the food and chemical industries to find new separation "green" techniques which typically use less solvent and energy, such as microwave extraction, supercritical fluid extraction, ultrasound extraction, ultrafiltration, flash distillation, the controlled pressure drop process and subcritical water extraction.

Ultrasound-assisted extraction is an emerging potential technology that can accelerate heat and mass transfer and has been successively used in extraction field. Ultrasound waves after interaction with subjected plant material alter its physical and chemical properties and their cavitational effect facilitates the release of extractable compounds and enhances the mass transport by disrupting the plant cell walls. Ultrasound is well known to have a significant effect on the rate of various processes in the chemical and food industry. Much attention has been given to the application of ultrasound for the extraction of natural products that typically needed hours or days to reach completion with conventional methods. Using ultrasound, full extractions can now be completed in minutes with high reproducibility, reducing the consumption of solvent, simplifying manipulation and work-up, giving higher purity of the final product, eliminating post-treatment of waste water and consuming only a fraction of the fossil energy normally needed for a conventional extraction method such as Soxhlet extraction, maceration or steam distillation.

21.4 The Use of Microwaves

Microwave heating is based on the transformation of alternating electromagnetic field energy into thermal energy by affecting the polar molecules of a material. The most important characteristic of microwave heating is volumetric heating. Conventional heating occurs by convection followed by conduction where heat must diffuse in from the surface of the material. Volumetric heating means that materials can absorb microwave energy directly and internally and convert it into heat. In microwave heating, heat is generated throughout the material, leading to faster heating rates, compared to conventional heating where heat is usually transferred from the surface to the interior. Microwave heating takes place because certain types of molecules are capable of absorbing electromagnetic radiation in the radio-frequency range. Water, the major constituent of most foods, is composed of such molecules.

Microwave drying is caused by water vapour pressure differences between interior and surface regions, which provide a driving force for moisture transfer. Microwave drying results in a high thermal efficiency, shorter drying time and improved product quality compared to conventional hot air drying. Microwave drying helps to remove the moisture from the food products without the problem of case hardening. Microwave hot air could greatly reduce the drying time of biological materials without damaging the quality attributes of the finished products.

Microwaves are highly penetrative and the absorption of radio-frequency energy within food causes a rapid rise in temperature throughout the material and thus produces a relatively uniform cooking effect. It is lucky that materials such as glass, ceramics plastics and paper, from which containers and packaging of foodstuffs are commonly made, do not absorb radio-frequency energy so that microwaves pass straight through them without making them hot. On the other hand, metals reflect microwaves and cannot be used in microwave cookers, although recent development in the design of ovens and packaging suggest that aluminium foil, which is a very important packaging material for frozen foods, may soon be able to be accommodated in the technique of microwave processing.

Microwave heating has positive ratings for drying rate, flexibility, colour, flavour, nutritional value, microbial stability, enzyme inactivation, rehydration capacity, crispiness and fresh-like appearance.

Two wave frequencies are commonly used: for domestic ovens, 2 450 MHz is preferred while, in industrial applications, 896 MHz is considered more suitable. The product to be heated is placed in an enclosed oven and microwaves, generated by special oscillator tubes, are guided onto it. The penetrative power of the radiation allows cooking to occur rapidly throughout the product, unlike conventional ovens, where cooking takes place from the outside of the food inwards and rates of heat transfer are relatively slow.

1. Domestic and Commercial Catering Applications

The most useful application of microwave cooking under domestic conditions and for those concerned with food service, has been in heating frozen cooked meals or snacks for consumption. A frozen meal, for example, can be heated to serving temperature in 1.5-4 minutes in a microwave oven compared with 20-30 minutes in a conventional oven. Where frozen foods are popular, the use of the microwave oven has grown rapidly. It has been estimated that in the United States, by 1980, 15% of all homes and 50% of all catering establishments had microwave ovens.

The cooking of fresh foods, particularly meats, in microwave ovens suffers from the disadvantage that microwaves cannot brown foods. This has to be done separately under a grill or in an oven and reduces the convenience of using microwaves. Attempts are being made to overcome this by incorporating within ovens a browning element or a special heat sink constructed of absorbing material on which the food may be placed. Beef roasts or poultry may be cooked in 6-7 minutes although this must be followed by a resting period of 15-20 minutes before carving to allow heat to penetrate evenly throughout the mass.

It is expected that the rapid growth in use of microwave ovens will force food manufacturers to develop products especially prepared and packaged for microwave cooking. New opportunities and challenges exist, in particular, for the frozen food industry and the first signs of these being taken up are already evident in the form of frozen pizzas, pancakes and popcorn.

2. Industrial Applications

The penetrative power of microwaves and their ability to raise temperatures rapidly infrozen foods has led to the widespread use of microwaves to assist thawing of frozen ingredient raw materials in the food industry. Meat, fish and poultry are normally used by the industry in the form of frozen 25–50 kg blocks. These have to be removed from frozen storage to be tempered at 0 ℃ for several days before use. In contrast, a 20 cm-thick block of frozen beef weighing 50 kg can be tempered from −15 ℃ to −4 ℃ in 2 min by microwave heating. It is then ready for processing. The commercial advantages of flexibility, energy-saving and the saving of refrigerated and frozen storage space are readily apparent. Frozen ingredients are not thawed completely. Because water absorbs energy more quickly than ice does, the first part of a block to thaw would get hot quickest so that before long water could be boiling at one place adjacent to another place which was still frozen.

Other industrial uses of microwave heating include:

(1) The production of skinless sausages;

(2) Porto crisp manufacture;

(3) An aid to freeze-drying and air-drying where case-hardening in the final stages of drying can be avoided;

(4) Baking of biscuits and cakes.

Although microwave heat treatment has many advantages compared to conventional methods, it is still not used widely for commercial purposes, which is due to both technical and cost factors. The quality of microwave-treated products is better or equal to that of conventional drying. However, there are certain drawbacks in microwave heating such as the non-uniform heat distribution and higher equipment costs. Equipment costs can change with time and developing technology. A major improvement in the efficiency of the treatment could change the economics of the microwave process. Thus, microwave heat treatment does appear to have a high potential for the processing of agricultural products in the near future.

Words/Phrases and Expressions

freezing 冷冻
cutting 切割
drying 干燥
tempering 回火
bleaching 漂白
sterilization 杀菌
extraction 提取
radio frequency 射频
Radio Frequency Identification (RFID) 射频识别
submergence 浸入
fryer 炸锅
crystallization freezer 结晶冷冻机
freezing by immersion 浸入冷冻

atomisation 雾化
hot gas stream 热气流粉碎
pulverisation 粉碎
marinatingbrine 腌制卤水
degassing 脱气,除去空气
filtration 过滤
filters 过滤器
membranes semipermeable 半透膜
demoulding 脱模
teflon moulds 特富龙模具
silicon moulds 硅模具
defoaming 消泡热处理
emulsification 乳化
oxidation 氧化
cavitation 气蚀
deaeration 脱气,脱泡
polarization 极化
cavitational 空化
sonicate 超声处理
penetrative 穿透的
flexibility 柔韧性
flavor 风味
nutritional value 营养价值
microbial stability 微生物稳定性
enzyme inactivation 酶灭活
rehydration capacity 补水能力
crispiness 脆性
fresh-like appearance 外观新鲜度
polytetrafluoroethylene 聚四氟乙烯
depolymerization 解聚
defrosting 除霜
thawing 解冻
tenderization 嫩化
miscellaneous 各种各样的

Chapter 22　Automatic Cooking Machine

22.1　Introduction

The cooking techniques used in the preparation of Chinese dishes were developed more than three thousand years ago. Chinese cooking technique is the most complicated of the three cuisine systems in the world (Chinese cuisine, French cuisine, Turkish cuisine). Chinese cuisine is one of the most complicated types of the world's cuisine systems. Chinese dishes have received a great deal of attention because of the attractiveness of their color, aroma, taste and appearance. The cooking processes in Chinese cuisine involve five main techniques: stir-frying, frying, boiling, grilling and steaming. Among these, stir-frying is the most representative of the features of Chinese cuisine. The cooking process of stir-frying is extraordinarily complicated. Traditionally all the tedious steps of the cooking process have to be done manually.

The cooking equipments available now can only make simple cooking processes such as heating with the microwave oven, baking with the roaster, boiling with an electric kettle, broiling with the frying pan. They can not complete the core part of the Chinese cooking processes such as pan-frying, stir-frying, burst-frying, quick-frying and re-frying without the related automated operations.

Chinese dishes are very famous, but their cuisine makes cooks tired. During the traditional cooking process, cooks often work hard in the hot kitchen. In the case, it is not surprising that the taste of most dishes is not as good as we desired. Furthermore, the high pressure of the modern life makes people spend less time in cooking, as well as the Chinese cuisine is very difficult. So it is urgent to design the machine for cooking to take the place of people.

The automatic cooking machine for Chinese dishes is a new machine with the following functions:

(1) It can put the raw materials into the wok automatically;
(2) It can make the food in the wok heated evenly;
(3) It can finish the basic Chinese cookingtechniques;
(4) It can clean itself.

It also should be easy to operate by people and the dishes that are cooked by the automatic cooking machine should be delicious. The Chinese cuisine has its own features, understanding the principle of Chinese cuisine is necessary for the design of the automatic cooking machine. In these days of increasing automation, there is a demand for an intelligent cooking device which can perform the complicated motions of a skilled chef. Now, with the development of robotic technology, a cooking robot which can prepare the delicious good quality dishes is in demand.

22.2　The Essential Cuisine Principle of Chinese Foods

The cooking method of Chinese foods is different from that of western foods. According to different cooking process, Chinese foods can be divided into five types. The cooking steps of each type are showed in Table 22.1.

Table 22.1　The steps of cooking for Chinese foods

No.	Type	The steps of cooking
1	Stir-fry dishes Series	Add the oil, add the major ingredients, add seasoning, pan-fry and stir-fry, add water and starch, stir, serve and finish
2	Braise and stew dishes Series	Add the oil, add the major ingredients, turn and stir-fry, add water, add seasoning, shakes the wok, serve and finish
3	Pan-fry and deep-fry dishes Series	Add the oil, add the major ingredients, shovel the ingredients, add seasoning, pour the oil, serve and finish
4	Saute dishes Series	Add the oil, add the major ingredients, turn the ingredients, add seasoning, pour oil, add seasoning juice, add water and ingredients, add the ingredients, stir, serve and finish
5	Soup Series	Add stock, add the major ingredients, add seasoning, take away the spume, add water and starch, drench the oil, serve and finish

By Table 22.1, we divide the actions of cooking Chinese foods into five parts: add the oil, add ingredients, pan-fry and stir-fry, stir, pour the oil or water. Where, pan-fry, stir-fry and stir are both achieved by pancake turner. However, the cooking machine is impossible to achieve different action by using different mechanism, if that, such mechanism will be extremely complicated. By synthesizing the characteristics of various actions, we can divide the process of cooking into four parts: feeding module, the module of leaving the material out in the middle process, cooking module and fire-controlling module. Each module can achieve some special functions. Table 22.2 shows the functions of each module.

Table 22.2 The main functions for different modules

No.	Module			Function
1	Feeding module			Add the ingredients into the wok accurately
2	The module of leaving the material out in the middle process	(1)	Pour	Get the ingredients out of the wok completely
		(2)	Drain the oil	Extract water and oil from the ingredients that treated after drained by oil
		(3)	Back to the wok	After the ingredients been drained of oil, they will be put back to the wok
3	Cooking module	(1)	Even stirring	Mix the ingredients and heat them evenly
		(2)	Stir-fry and disperse	Separate the ingredients immediately and heat evenly
		(3)	Gathers together	Gather the ingredients into the bottom of the wok
		(4)	Turn another side	invert the ingredients completely (Heating surface and relative non-heating surface inversion)
		(5)	Guards against sticks	Prevent the conglutination of the ingredients with the wok in order not to be burnt
		(6)	Crack	Crack the ingredients conglomerated by heating into pieces (separate)
		(7)	Drench	Let the stock cover or soak the ingredients
4	Fire-control module			Control the switch of cooking range and the size of fire

The materials of the dishes are arranged according to the traditional Chinesetechniques, and the motion information every dishes are produced by Chinese cuisine experts using the robotic method of "teach-in and playback". Based on the essential Cuisine Principle of Chinese Foods we can design a new cooking machine system for Chinese foods.

22.3 The System of the Chinese Foods Cooking Machine

22.3.1 The Essential Mechanisms of the System

The cook is playing the primary role in the cooking process that is finished by manual work. He can decide when to add the ingredients, how much to add and how to operate in order to make the food taste better. However, the machine is limited in flexibility, it is impossible to imitate people cooking completely. It must coordinate various movements of mechanisms in order tofulfill the requirement of manual cooking.

By Table 22.2, the structure of the Chinese cooking machine consists of four essential parts: feeding mechanism, the mechanism of leaving the material out in the middle process, wok movement mechanism and stirring-fry and dispersing mechanism. The cooperation of these four parts can fulfill the movement showed in Table 22.2. In other words, the feeding mechanism can put the right ingredients into the wok accurately. The wok movement mechanism and stirring-fry and dispersing mechanism can accomplish nearly all the cooking methods, and can drain the oil with the mechanism of leaving the material in the middle process. The stirring-fry and dispersing mechanism is propitious to the cooking of soup. At the mean time, the automatic cleaning mechanism will be designed into the stirring-fry and dispersing mechanism. The fire-control system is closely related to the wok movement mechanism, the fire-control should be performed in the whole cooking process.

1. Wok Movement Mechanism

The wok movement mechanism is the core mechanism of the automatic cooking machine. The wok is shook and leaned by hand, when this work is finished manually. So the design of the wok movement mechanism is the key for the machine. The functions are shown in the following:

(1) Heat the ingredients evenly;

(2) Cooperate with the stir-fry and dispersing mechanism;

(3) Cooperate with the mechanism of leaving material out in the middle process;

(4) Make the agglomerate ingredients in the wok turn over fully or partly;

(5) Pour the used oil or water into the oil tank or waste trough.

Figure 22.1 shows the schematic diagram of the wok movement mechanism, which can achieve three different movements: level moving, shaking and turning. These movements can implement the actions bellow.

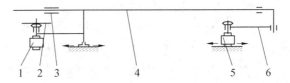

1-motor1; 2-conic-gear1; 3-conic-gear2; 4-rocker; 5-motor2; 6-rank

Figure 22.1 The principle of the wok mechanism

(1) Level moving can move the wok center. In different positions, the wok can cooperate with others to add ingredients and put the material out in the middle process.

(2) Shaking the wok is the basic movement of Chinese Cuisine, which can make the food heated evenly, and also can fully turn over the agglomerate ingredients in the wok by changing the velocity and acceleration. The velocity of shaking wok should not be too fast, or the ingredients may be thrown out.

(3) Turning movement not only can throw the oil or waste water out of the wok, but also can help to pour and accelerate the turning of the ingredients by the mechanism of leaving the material in the middle process, in order to heat the ingredients evenly.

2. The Stirring-Fry and Dispersing Mechanism

This mechanism plays the important role in the process of cooking Chinese foods. It is composed of comb-shaped stirrer used for stirring-frying and dispersing and the cover used for gathering together (Figure 22.2). In order to simplify the entire mechanism, we should install the nozzles inside of the cover to achieve the function of cleaning.

(1) To save water, six small nozzles are separately installed in the cover of the wok in different positions, so that the wok can be cleaned with less water.

(2) The function of stirring is to make the sticky ingredients heated evenly, so that is prevented the phenomenon of dishes sticking to the wok.

Figure 22.2 The 3D drawing of the cover mechanism

3. Analysis of Featured Motions in the Cooking Process

In the cooking process of stir-frying, a series of cooking motions are undertaken, such as pushing, pulling, raising, rocking, lifting, pouring, turning, etc.. These cooking motions enable the material being cooked to keep moving relative to the wok, this will prevent the material from sticking to the wok, and also allow the food to heat evenly, so that a high quality of color, aroma, taste and appearance can be guaranteed. Four main motions are featured in the wok motion mechanism of the cooking robot: rocking, turning over, tilting and moving the wok as shown in Figure 22.3. The details of the four featured motions are given as follows.

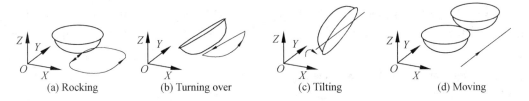

Figure 22.3 Featured motions of cooking process for stir-frying

(1) Rocking. Rocking is realized through two translations along the directions X and Y or by only one translation along the direction X or Y, as shown in Figure 22.3(a). This causes the wok to move in an approximately circular motion or linear reciprocating motion. These motions will prevent the food being cooked from adhering to the wok, it also helps it to blend evenly with the cornstarch. The cornstarch input is an important cooking step for most of the stir-

fried dishes. Rocking can be classified into circular rocking and linear rocking.

(2) Turning over. Turning over is the core cooking process for stir-fried dishes, it is fulfilled through two translations along the directions of Y, Z (or X, Z) and one rotation around X (or Y) axis, as shown in Figure 22.3(b). When turning over, the wok performs an approximately parabolic motion. Turning over enables the material being cooked to be uniformly heated so that it is consistently mature and of an even uniform color.

(3) Tilting. The tilting motion requires one rotation around the X (or Y) axis as shown in Figure 22.3(c). The aim of tilting of the wok is to toss the cooked material up into the air, out of the wok, and catch it neatly in its original position maintaining the original position of the wok after the toss. The tilting motion happens when the wok is inclined in order to toss the cooked material.

(4) Moving. The moving motion demands one translation along the direction of X (or Y) as shown in Figure 22.3(d). The aim of this movement is to make the wok move from the heat source before tossing the cooked material and moving the wok back to the heat source after tossing the material. Moving consists of moving away and moving back.

Conceptual design of the wok motion mechanism is as follows. Based on the requirements of the featured motions for the cooking process, a 4-DOF hybrid serial/parallel manipulator including a 3-DOF parallel manipulator plus 1-DOF translation is suitable to be the wok motion mechanism of the cooking robot. The 1-DOF translation can be simply realized by a linear guide, the main focus is on the conceptual design of the 3-DOF parallel manipulator with 2-DOF translations and 1-DOF rotation.

As is well known, the cooking process and temperature control are important factors in the traditional cooking techniques of Chinese cuisine. An example of a typical cooking robot of 2T (2 - Translations) configurations, pure 1R (1 - Rotations) configurations is shown in Figure 22.4. From the figure, it can seen that their possessing fewer active legs, having a simple structure, being easy to control and providing easy kinematic solutions. During the cooking process the end-effector of the cooking robot needs to implement 2-DOF translations in a plane and 1-DOF rotation perpendicular to the plane or parallel to one axis of the plane, so both planar and spatial 2T1R parallel architectures are highlighted in the wok motion mechanism of the cooking robot.

There are various kinds of 3-DOF planar parallel manipulators with non-identical limbs. The 2T1R 3-DOF parallel manipulator can stem from the five-bar mechanism as shown in Figure 22.5. Two bars (bars AB and DE) serve as the input bar. If the bar CD acts as the moving platform of a parallel manipulator, the platform will have two d translations 2T (Y, Z). When the revolute pairs in the pivots (C and D) are substituted by a universal pair or a spherical pair, as shown in Figure 22.5(b) and (c), the moving platform (bar CD) possesses a local rotation R(Y) which rotates around the axis CD.

It is obvious that the moving platform has two translations and one rotation when the bar

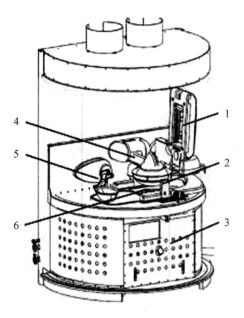

1-Material feed module; 2-Pan motion module; 3-Robot control module
4-Stir-tool module; 5-Secondary motion module; 6-Fire control module

Figure 22.4 Diagram of a cooking robot

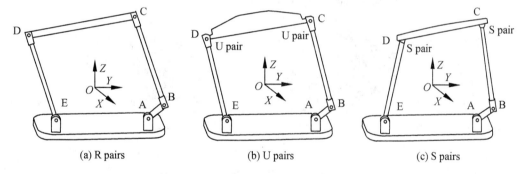

(a) R pairs (b) U pairs (c) S pairs

Figure 22.5 Five-bar mechanisms composed of R, U and S pairs

CD is enlarged to a plane. If the third limb with 6-DOF is attached to the moving platform, the DOFs of the moving platform will not change. So a family of novel 2T1R 3-DOF parallel manipulators is obtained as shown in Figure 22.6. The schemes of the wok motion mechanism for the cooking robot are given in Figure 22.7(a) and (b); based on Figure 22.7(a) the schemes of the wok motion mechanism for the cooking robot are given in Figure 22.7(c) and (d).

As shown in Figure 22.7, the 2T1R 3-DOF parallel manipulator is composed of R pairs, U pairs or S pairs. The scheme with only R pairs is more suitable to be used in the cooking robot because of its fast response and ease with which it can be driven. Therefore, the scheme in Figure 22.8 is developed as the wok motion mechanism of the cooking robot.

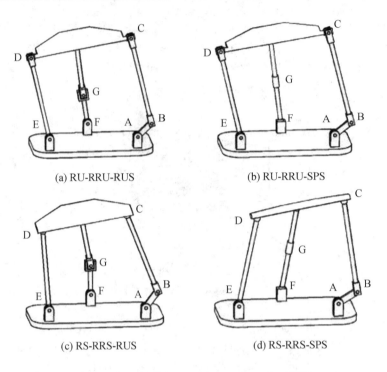

Figure 22.6 Some parallel manipulators generating 2T1R motions

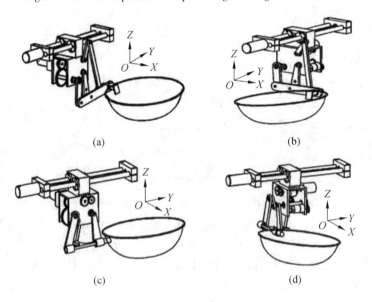

Figure 22.7 Schemes of the wok motion mechanism for a cooking robot

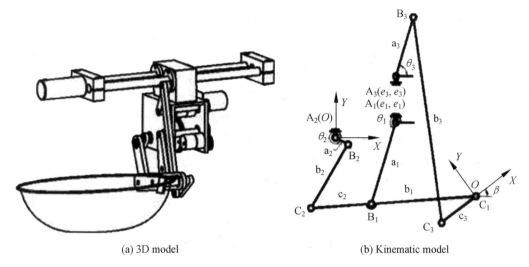

(a) 3D model　　　　　　　　　　　(b) Kinematic model

Figure 22.8　A novel 4-DOF parallel manipulator

4. The Feeding Mechanism

For automatic cooking machine, there will be a standard menu in which the kinds and rates of ingredients have been determined, so this mechanism just needs to send the various ingredients into the wok accurately at the pre-arranged time. We can divide the feeding box into 4 parts, and keeping oil, major ingredients, and minor ingredients and seasoning separately in each parts, then seal them with the film. The ingredients will drop into the wok which has already been moved accurately under it. The function of feeding ingredients is automatically achieved.

5. The Mechanism of Leaving the Material in the Middle Process

This mechanism will cooperate with the wok movement mechanism to achieve following motions when deep-fried food: pour out of the wok, drain the oil and bake the wok. This mechanism needs two motions: move up and down and turn over.

22.3.2　The Control System of Automatic Cooking Machine

As a household appliance, the automatic cooking machine should be convenient and reliable. The method of modularization and object oriented processing could be used in the software of the machine. It mainly includes monitor module, movement module, fire-control module and the human machine interface (HMI). As the former two parts need to exchange information, we adopted the CAN bus communication module to ensure the stability and reliability. The monitor module is an important part of the automatic cooking machine. This module includes the functions as follows.

(1) Identify the bar coded of box semi-manufactured ingredients produced by vegetable preparing center to affirm the name of the dish to be cooked.

(2) Environment monitoring (the temperature, the density of CO in air and so on).

22.3.3 The Fire Control System

The fire control is the key and also the difficult points for automatic cooking machine to imitate the real cooking methods. It aims at achieving intelligent step-less nonlinear control of the cooking-power according to different dishes, flow-rate of flammable gases in different area. The fire control unit consists of the micro-processor controlling part, the action performing part, the temperature closed-loop part and the safety monitoring part.

1. The Micro-Processor Controlling

The tasks of signal processing, the temperature closed-loop control, safety monitoring and the control of different valves require wide temperature adaptation and high operation speed. The corresponding real-time operation system is valuable to insure the accuracy, consequently implement multi-tasks parallel processing and promote the operation efficiency.

2. Action Performing

The executing objects of the entire cooking-power auto-control system includes: cooking-power intensity valve control, security valve control and ignition coil control and so on. Cooking-power intensity control is the key of the entire cooking-power controlling system. The flow-rate can be regulated through mechanic or electric methods. Together with temperature and thermal radiation real-time monitoring the autonomous closed-loop feedback cooperative control can be achieved. The security valve control is the important part for the cooking-power control. As the complete cooking-power control unit should achieve the gas appliances safety certification. So when the cooking-power intensity controlling part can not fully cut down the gas, the security valve, using relay or power device to perform reliable control, should perform the shutdown action to insure the safety of the system.

3. Temperature Closed-Loop

Temperature closed-loop control can be achieved by detecting temperature at fixed positions. Because wide temperature adaptation is required, reliable temperature could be measured by the thermal radiation sensor and platinum resistance by means of difference amplifying and nonlinear revising. Temperature measured at fixed positions is not only a reference for the cooking-power control, but also a feedback of the control effect.

4. The Safety Monitoring

As it concerns the flow-rate control of flammable gases and gas appliances should reach the safety certification standards, so the security of the system is especially important. CO concentration sensors are used for safety monitoring, and the flow-rate of flammable gas is controlled by the security valve. Safety alarm can be made by the cooking-power intensity controlling system and the main controlling system, which require timely maintenance.

22.3.4 Cooking with Robots Working in Open Environments

Cooking in a closed environment consumes too much space for typical home kitchens.

When using a system that works in an open environment, the user puts preprocessed cooking ingredients on the table and has small robots execute the cooking tasks using a pot on an induction heating (IH) cooker. When not using the system, the user can cook on the same table using the same cooker.

First, small robots are used in place of built-in arms to save space and increase safety. They are also designed to use common cooking utensils. Second, small tags with visual markers could be used for the robots and object tracking in the environment. The user can easily add and remove these markers to associate real objects with virtual information in the system. Third, a graphical user interface should be provided for giving cooking instructions to the robot. The user can control the robots, by instructing them when to add each ingredient to the pot, and the strength of the cooker.

A graphical user interface is a better method for giving cooking instructions to robots. A state-of-the-art humanoid robot has demonstrated partial cooking tasks. However, the user has to write a program for instruction. A "cooking with robots" realizes with a graphical interface.

1. A Household System Overview

Figure 22.9 shows an overview of our system. Several small mobile robots on a customized table add ingredients and seasonings to a pot on an IH cooker. The ingredients are placed on customized plates, and the seasonings are in customized bottles so that a robot can handle them. The location of the robots, plates, and bottles, are tracked using visual markers attached to them and with a ceiling camera. The robots, the IH cooker, and the camera are all connected to a controlling PC.

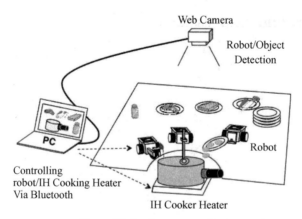

Figure 22.9 System overview

The user starts cooking either by selecting a desired recipe in the recipe book or defining a new recipe. The user presses the start button to start the system cooking. The system then puts the ingredients into the pot one by one and adjusts the heater strength according to the predefined procedure and notifies the user upon completion. The user first preprocesses the necessary ingredients, places them and the appropriate visual markers on the plates. The system comes with a set of visual markers for common ingredients (onions, carrots, potatoes, beef, pork, etc.) but user can also use supplemental visual markers for other ingredients. User then defines the cooking procedure, which is, what ingredients to use and how to adjust the heat using a graphical user interface.

2. Implementation

(1) The environment.

A special table placed higher than the pot so that the robots can easily pour ingredients and seasonings. The plate has handles around it for grabbing and a pole at the center for placing markers. The IH cooker is implanted a micro controller circuit and Bluetooth communication module inside for controlling the temperature level remotely.

(2) Mobile robots.

There are three types of customized small mobile robots, one for transporting the ingredients on plates, one for transporting seasonings in bottles, and one for stirring the pot. The first robot grabs a plate using a single arm, moves to the pot, and tilts the plate to drop the ingredients into the pot. The second robot grabs a bottle by using a hand with two fingers, moves to the pot, and shakes the bottle to sprinkle the seasoning into the pot. The amount of the seasoning is specified on the interface by dragging a seasoning icon multiple times. The stirring robot stirs the pot at appropriate times. Before cooking begins, the user needs to select and attach a cooking utensil to the robot. These robots do not have any sensors and are wirelessly (Bluetooth) controlled with the control computer. Small mobile robots carrying plates and bottles are more appropriate than built-in arms and containers because a user can store them in a cabinet when user cooks in the same environment without using the system.

Words/Phrases and Expressions

stir-fry 炒
fry 炸
boil 煮
grill 烤
steam 蒸
pan-fry 煎
stir-fry 翻炒
quick-fry 爆炒
burst-fry 炸
re-fry 重炸
rock 摇摆
turn over 翻转
tilt 倾斜
move 移动
parallel manipulator 并联机械手

Chapter 23   Design of Robot for the Food Industry

23.1   Introduction

  The food industry is under growing pressure to reduce production costs due to high levels of competition between supermarket chains at home and rival competitors abroad, where the cost of labour is significantly lower. The industry is now looking increasingly towards automation and robotics to help lower production costs further. Currently, however, the majority of industrial robots are largely unsuitable for the specific needs of food production. While the use of robots for end-of-line operations such as case packing and palletising is already well established, robotic automation of primary packaging and assembly of foods has so far been limited. Financial justification for the installation of a robotic system is typically based on the reduction of labour costs. The bulk of manual labour in a food production line is generally concentrated in primary packaging and assembly operations, which require high throughput, increased levels of hygiene and greater flexibility. Fixed automation machinery is sometimes used on dedicated production lines and although this is generally low cost to build once designed, costs for development and time invested can be very high. In a plant with many unique automatable tasks, it may be more cost effective to reduce labour costs by using low-cost robotic manipulators that can be quickly deployed next to the conveyor belt and rapidly reprogrammed to start on a new task alongside human operators. These robots would be used to automate simple tasks for which special purpose machinery would otherwise have to be developed.

  To date very few commercial robots have been developed specifically for the food industry. Often, existing models have simply been upgraded for use in food production. Nevertheless, in recent years, a small number of new industrial robots targeted at the food sector have begun to appear. For example, the KUKA KR 15 SL is an all stainless steel robot that has been used in the brewing and dairy industries. Some of the most recent robots to appear on the market include the Adept Quattro s650HS, ABB IRB 360 FlexPicker, FANUC M-3iA and the Motoman MPK2. The chapter is to examine what impact the unique requirements of the food industry have on robotic manipulator design. In order to design a robotic arm for use in primary packaging or assembly of foods, it is vital to keep these functional requirements in mind from the very outset of the design process. The current need for a robotic arm specifically for use in the food industry has led to the development of a manipulator. The manipulator exhibits very good design characteristics in terms of hygiene, high speed and low cost.

23.2 Design of Food Industry Robot

This section gives an overview of the design features of a low cost industrial robot prototype that has been specifically designed and developed for use in the food industry. The robot has been designed to meet the performance specifications set out. The design criteria are discussed in general terms and can be equally applied in the development of other food grade industrial robots. Perhaps the largest number of constraints imposed on manipulator design stem from the requirement for ease of cleaning. However, the application of hygienic design principles will often result in a simple, robust design. For the robot to be successful commercially, and gain industrial acceptance, requires that emphasis be placed on delivering optimal performance in terms of hygiene, speed and reliability at low cost.

23.2.1 Configuration

The choice of configuration for the robot is governed by the requirement for an easy to clean, sealed housing. Rotary joints are relatively easy to seal effectively. Linear joints are much more difficult to seal. The reciprocating motion of a linear joint tends to draw food particles inside of the joint, where micro-organisms can survive cleaning and multiply to later re-emerge and contaminate the product. For this reason, all of the robot's joints should be rotary joints. The use of bellows to seal linear joints with strokes longer than a few centimetres should be avoided as these are generally difficult to clean. The SCARA configuration, for example, makes use of a linear joint which is difficult to seal. Other configurations with all rotary joints are limited to articulated arm and Delta robots. Although all the driven joints in a Delta robot are rotary, the delta structure employs universal joints that are difficult to clean and ball and socket joints that require disassembly for cleaning. The articulated arm configuration has a small number of easy to seal joints and is well suited for hygienic design.

23.2.2 Degrees of Freedom

With the exception of a few specialised applications, the vast majority of tasks in the food industry are planar tasks performed on a conveyor belt and can therefore be accomplished using four coordinated degrees of freedom. The use of a four, rather than a six, axis manipulator has several advantages. Fewer motors and associated controllers are required, leading to a substantial reduction not only in cost, but also in the mass and inertia of the robot. The kinematic and dynamic equations of the manipulator are greatly simplified. Additionally, a six degree of freedom coordinate system is very difficult and confusing for humans to visualise. The use of a four degree of freedom coordinate system greatly simplifies user programming of robot motions. If required, an additional uncoordinated degree of freedom can be built into the robot gripper for tasks where the object being handled must be tilted out of the plane.

23.2.3 Reach and Workspace

Conveyor belts used in the food industry are generally no more than 50 cm or 60 cm wide. This width corresponds to the maximum distance that an operator, standing on one side of the conveyor, can comfortably reach across to pick an object on the opposite side. Similarly, the robot will need to be able to reach across the full width of the conveyor to pick objects both on the near and far sides. Although an articulated arm robot has a large work envelope, the vast majority of tasks will be carried out in a relatively small region over the conveyor belt. The robot is required to move horizontally at high speed (80–100 picks/min) within this region for pick and place tasks, but much lower speeds would be acceptable in other parts of the work envelope. Therefore, rather than specifying that the robot be capable of attaining high velocities throughout its work envelope, it is instead optimised for high pick and place speed only in the region directly above the conveyor. This also simplifies the design process to some extent by reducing the region of the work envelope that must be analysed through mathematical modelling of the manipulator dynamics. The robot may additionally be required to work in a ceiling mounted configuration. It is therefore designed so that it will have similar performance whether floor or ceiling mounted and it is possible to change from one configuration to the other relatively easily.

23.2.4 Links

The design of the robot links must fulfil several important requirements. The links should be lightweight, to minimise the inertia of the arm, low cost to manufacture and durable to last for the lifetime of the robot. This should be around seven years at least (approximately 40 000 hours of operation) and is ultimately determined by the life of the transmission bearings. The links must have an easy to clean, hygienic design and provide a sealed, waterproof enclosure to protect internal components. Additionally, the links should be stiff enough not to deflect significantly under load, in order to maintain accuracy and also be able to bear the stresses exerted on them by the transmission.

Industrial food processing machinery is almost universally manufactured from stainless steel. This preference is due to the fact that in the food industry hygiene is considered to be the most important factor for material selection. Robot designers have often looked to alternative materials, surface coatings or external coverings to provide advantages in terms of reduced manufacturing cost or decreased mass. However, for the food manufacturer these represent a real disadvantage if the robot is to be used in an application requiring high levels of hygiene. The use of plastics and composites in some industrial robots has allowed for an increase in speed through the reduction of link mass. Plastics have a tendency to creep under sustained load and are therefore not suitable for the manufacture of the links, unless an internal supporting metal structure is used. Carbon fibre is perhaps the only material with sufficient resistance to bending, creep and corrosion to provide a viable lightweight alternative to stainless steel. Despite its low weight, the use of carbon fibre has some disadvantages in terms of hygiene as discussed previously. The challenge therefore remains how to build a stainless steel robot at low cost.

23.2.5 Joints and Seals

Careful attention should be given to the design of the robot joints to ensure that they are both waterproof and hygienic. In many industrial robots, there are deep, narrow crevices at the joints which are impossible to clean. This is illustrated in Figure 23.1(a). An improved design using a spring-energised PTFE face seal is shown in Figure 23.1(b). Commercial sanitary seals are available where the spring is completely enclosed within the seal for use in food processing applications. Cover plates, providing access to the inside of the robot, are sealed using rubber gaskets. The screws securing the cover plates should also be sealed. Small screws can be sealed using a food grade sealant. The screws should have plain hexagon heads, which are easier to clean. The joints are designed to provide separation between the links, allowing sufficient space for visual inspection and manual cleaning around the joints. A detailed view of the robot elbow joint is shown in Figure 23.2.

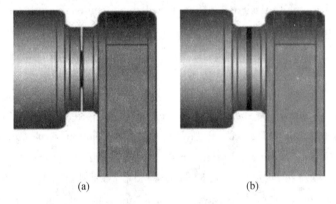

Figure 23.1 Unhygienic and hygienic robot joint design

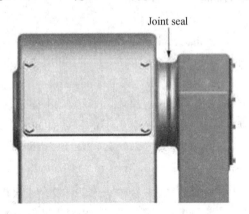

Figure 23.2 Detailed view of robot joint 3 (elbow) showing hygienic joint sea

23.2.6 Actuators

Pneumatic cylinders are low cost and commonly used in the food industry to actuate fixed automation machinery. Accurate position control of pneumatic actuators without the use of mechanical stops is, however, difficult to achieve under either proportional or pulse width modulation control schemes. Additionally, the solenoid valves used to drive the pneumatic actuators typically have a finite life of a few million switching cycles and would require frequent replacement if used in high speed pick and place robots, increasing maintenance costs and downtime. Hydraulic actuators are not used in the food industry, as there is a risk that hydraulic fluid may contaminate the product. Electric motors are comparatively easy to control and reliable in service. Brushless DC servomotors, despite their higher initial cost, have a service life many times greater than that of brushed motors, making them more economical to use over the lifetime of the robot. AC servomotors are generally used in commercial industrial robots. A wide range of commercial servomotors are available and these can be ordered preassembled with suitable gearheads, brakes and encoders. For high speed applications the selection of an appropriate motor and gearhead combination for each joint is crucial. The selection procedure should include a comprehensive dynamic simulation of the manipulator. The motors are controlled using off-the-shelf industrial controllers. A simple individual joint PID control scheme is used, as is generally done on most present-day industrial manipulators.

23.2.7 Transmission

An illustration of the robot transmission can be seen in Figure 23.3 and a second partial view is shown in Figure 23.4. Timing belts are used to transfer power from the motors to each of the joints. Standard length polyurethane belts are used. Rubber belts should be avoided as they shed particles as they wear. Pulleys are designed so that they can be machined from toothed bar stock and made using a lead-free aluminium alloy. Four brushless DC servomotors are used to drive the robot joints. The joint 1 motor (first shoulder joint) is housed in the base. Joint 2 (second shoulder joint) and joint 3 (elbow) motors are housed inside the shoulder close to the robot base, significantly reducing the inertia of the arm. The use of this arrangement results in a significant dynamic de-coupling of joints 2 and 3. This is also particularly beneficial, as the resulting kinematics reduces the velocity required from the joint 3 motor for a given straight line motion (compared to that which would be necessary if the motors had driven the joints directly). The result is that faster operational pick and place speeds can be attained. Although the arm requires only four degrees of freedom, the requirement for all the joints to be rotary necessarily results in a robot with five joints. The first wrist axis (joint 4) must be oriented such that the second wrist axis (joint 5) always remains vertical. In order to avoid having an additional motor to control joint 4, this joint is instead linked via a series of pulleys to a fixed shaft in the shoulder. This fixed shaft keeps the orientation of the joint 5 axis constant.

A similar arrangement can be observed in many palletising robots, where external links are instead used to orient the wrist. The joint 5 motor is relatively small and is best located inside the robot wrist, driving the joint directly. Planetary gearheads are used as they are more easily backdriving. Part of the total gearing ratio required for each joint is supplied by the pulley transmission, permitting the use of smaller, lighter and less expensive gearheads. Motor torque is transmitted through the centre of each joint via a system of concentric shafts. A central hole though each joint allows for routing of wires and pneumatic tubing. Standard ball bearings are used in the transmission to minimise friction. These are sized to last for the lifetime of the robot without requiring replacement. To maintain accuracy, care must be taken to minimise sources of backlash in the system. The transmission should be designed for ease of assembly and maintenance. The use of stock parts means that spares will be readily available. During normal operation and over the course of its lifetime, the robot will be subjected to hundreds of millions of repetitive stress cycles. Owing to limitations on weight and space, transmission components will be relatively small and therefore highly stressed. Under these conditions, fatigue failure becomes a serious concern. Parts should consequently be designed to keep stresses below levels that could lead to premature failure through fatigue.

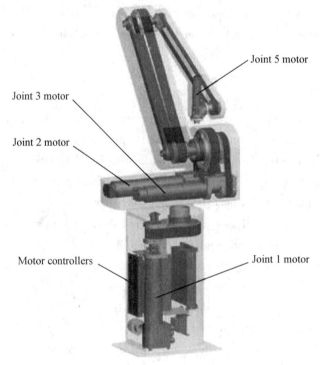

Figure 23.3　Illustration of the robot transmission and internal components

Figure 23.4　View of transmission components housed in the robot shoulder

23.2.8　Motion Control

To be truly successful in the food industry not only must a robot be hygienically designed, but it should also be fast. The joint torques required to move the robot through a particular path are highly dependent on how the motion is carried out. Slight differences in the shape of the path and the way that the displacement, velocity and acceleration vary as the robot travels along that path can result in surprisingly large differences in the torques that must be generated by the motors. Exactly how fast the robot will be able to move will depend on a number of constraints, such as the stresses that the transmission components can bear, limits on the torque that the motors are able to deliver without overheating, the maximum rated speed of the gearing and also the voltage and current that the power supply can provide. Motion control is used to maximise pick and place speed by means of an optimised trajectory planning algorithm. In addition, asymmetric acceleration profiles are used to further increase the robot's pick and place speed, especially near the edges of the work envelope. By accelerating asymmetrically, for example, having rapid acceleration at the start of the motion followed by gentle deceleration at the end (or vice versa), the motion can be optimised so as to minimise peak joint torques. This reduction in peak joint torques potentially allows for an increase in speed if all other constraints mentioned above are met. The degree of asymmetry to use for a particular trajectory is position dependent, so precomputed values are stored in a look-up table. Suitable values are found through inverse dynamics modelling of the robot arm.

23.2.9 Programming

Currently, the robot is programmed through a programmable logic controller (PLC). Future work may, however, involve the implementation of more intuitive and user-friendly methods of programming, such as programming by demonstration. For the moment, the use of a PLC offers several advantages. PLCs are industry standard equipment commonly found in most food factories. Consequently, the robot will be able to communicate directly with other equipment in the factory as well as network with other food robots, or even be connected to a central SCADA system. Sensors, vision systems (smart cameras) and conveyor encoders for object tracking can all be connected to the PLC either directly or via the network. Any discrepancy could signal a possible collision or malfunction, in which case the control system would act to bring the robot safely to a halt and shut off power to the motors. The robot can be confined to operate in a specified region of its work envelope using user-defined virtual safety barriers (defined in software), which the robot is not allowed to cross. These can be used in conjunction with physical safety barriers as a redundant safety mechanism or to avoid collisions with surrounding equipment.

23.3 Robots in Food Industry

Despite a complex supply chain which extends from processing, through packaging to storage, transport, distribution and retail, the food industry is not a major user of robots, with annual sales presently representing only around 2-3 percent of the total market. However, the industry is facing mounting pressures to adopt greater levels of automation, notably as a consequence of the difficulty in attracting and retaining staff to conduct what are often physically demanding and repetitive tasks, combined with high-food prices in an ever more competitive environment, leading to the need to minimise costs and optimise efficiency. This paper reviews existing uses of robots in the food industry and considers a number of new and emerging applications.

23.3.1 Robots in Food Processing and Packaging Food Industry

The food processing and packaging sector makes significant use of robot. Picking, packing and palletising are the leading food industry applications and typically involve two-axis packaging, four-axis palletising and six-axis palletising and handling robots. All manner of products are handled in this manner, ranging from unpackaged foodstuffs such as cheese, meat, poultry and fish, to cartons, drums, bottles, trays, boxes and other food and beverage containers (Figure 23.5). Although these are conventional robot applications which differ little from similar uses in other industries, hygienic design and/or high-operating speeds are often critical and several robot manufacturers have recently launched products aimed specifically at this market.

Most now feature an IP-67 rating which ensures complete protection against any ingress of dust (specified by the first digit) and the ability to be completely immersed in water with no ingress (second digit). Many food industry robots also have smooth external finishes which eliminate retention areas for contaminants such as food particles or bacteria. In addition, some new articulated designs feature a completely hollow arm that encloses the air and electrical utility lines. This type of sealed construction can withstand acidic and alkaline food industry cleaners as well as allowing wash-down with pressurised water without the need for disassembly.

The speed issue has long been problematic and the "spider" or Delta-style robots, developed in the early 1980s by Raymond Clavel at the Swiss, initially became popular for high-speed picking and placing of light objects (1 kg) such as chocolate bars. At that time, they were the only sufficiently fast robots available other than selective compliant assembly robot arms. Spider speeds range from 90 cycles/min to 150 cycles/min (a cycle is defined as picking an object, raising it 25 mm, moving it 300 mm, lowering it 25 mm to place it, then returning in the same pattern to pick the next object). Articulated robots for high-speed applications appeared more recently. For example, Fanuc introduced its five-axis, M-430iA/2 F (Figure 23.6) which is articulated and optimised for high-speed picking and packing, being capable of operating at up to 120 cycles/min with a 1 kg payload, while KUKA launched its Food Series options. These feature a robot arm with IP protection, food-grade oil in the robot reducers, a USDA-certifiable white coating to withstand common industry sanitisers and a stainless steel control cabinet.

Figure 23.5 A typical food industry palletising robot

A growing food processing application is robotic meat cutting. The use of robots in what

Figure 23.6　Fanuc's M-430iA/2 F robot

has traditionally been a skilled and labour-intensive task reflects both the difficulties in retaining staff and the inconsistencies arising from manual cutting. By way of an example, Mitchell's Gourmet Foods, a Saskatoon, Saskatchewan-based slaughterhouse, has installed a Fanuc M-710i robot, equipped with a vision system, to process 700 hogs or 1 400 hog bellies/h, a production rate that formerly required a six-man crew. The entire cutting process takes only 2.5 s and cutting is achieved with double-sided Denver knives which can be changed automatically. The economic benefits of robotic handling and processing are well illustrated by an installation at a Spanish lettuce processing plant owned by El Dulze. Lettuces taken straight from the ground are delivered to the plant in plastic containers stacked on a euro pallet. This system has reduced labour costs by 80% and damage to the lettuces caused by human handling is minimal, with reject rates falling from 20% to 5%.

23.3.2　Robots in Retail

Several robot manufacturers have argued that the food and drink retail sector offers significant future opportunities and although the market is still in its infancy, a number of products have recently appeared. Whether these are mere gimmicks or genuine steps toward true retail robots remains unclear but a taste of things to come is provided by Motoman's "RoboBar" product line (Figure 23.7). This is based on a standard dual-arm robot and is available in high-production (HP), entertainment (E) and non-alcohol (NA) versions, each designed to address a particular market niche. Motoman has sold an NA version to a company in Dubai where it is being installed in a futuristic office building, whilst the E model pours cocktails directly from bottles and one is in the process of being installed at the Harrods Department Store in London. Another retail innovation is Discovery Ice Cream's robotic ice cream vending machine (Figure 23.8). This American company uses a small five-axis robot arm to manipulate a scoop between various stations to dispense measured amounts of ice cream and toppings. A number of innovative food preparing and vending robots were shown. Okonomiyaki (Japanese pancake) robot (Figure 23.9) that is based on Motoman components and able to make pancakes automatically, covering the entire process from mixing the dough to flipping, and is the brain-child of robot system integrator Toyo Riki Engineering. The robot relies on speech recognition technology to take verbal orders from customers. Another was the "Chef Robot" (Figure 23.10), based on a Fanuc M-430iA and "H", a hand-type robot developed by Tokyo-based Scuse, which is able to prepare sushi.

Figure 23.7　Motoman's RoboBar

Figure 23.8　Robotic ice cream vending machine

Figure 23.9　The robot pancake maker from Toyo Riki

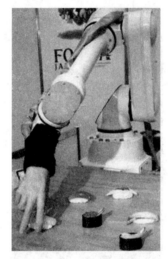

Figure 23.10　The sushi-making Chef Robot

23.4　Conclusion

A robot designed for the food industry must be able to fulfil a series of very specific requirements with regards to hygiene, cost, operational speed, safety and ease of programming. These are different from those of most industrial robots, which have been developed for generic use over a wide range of industries.

The use of robotic automation in the food industry, however, is extremely limited and lags behind other labour-intensive endeavours such as the manufacture of vehicles or electronic goods. The leading applications are picking, packing and palletising, together with niche uses in farming and the processing and retail sectors. This reflects several factors, particularly the technological difficulties in emulating the human perceptual and dextrous capabilities required to, for example, decide when a crop is ready for harvesting or make a sandwich. Further, unlike PCBs or car parts, foodstuffs are often very variable, soft and not easily handled and a significant proportion of companies are SMEs who lack the knowledge or funds to consider robotics. Improved robot designs aimed at meeting specific food processing needs are emerging and the falling prices of sophisticated vision systems, combined with improved grippers and user-friendly software are gradually resolving these problems.

For a robot design to be commercially successful and gain industrial acceptance in the food manufacturing sector, these issues must be addressed from the very outset of the design process. Food industry requirements for manipulator design have been reviewed and discussed in detail. The design of a low-cost robotic arm for the food industry, specifically developed with these requirements in mind, is also presented. Designed to be capable of operational pick and place speeds comparable to those of a human operator, these robots would be used to automate simple tasks in primary packaging and assembly of foods for which special purpose machinery would otherwise have to be developed. Future work will involve testing of the robot prototype in a real industrial environment and evaluation of its performance. The software and programming interface for the robot will continue to be developed and more capabilities added, such as programming by demonstration. Future design modifications may also include the addition of active compliance control of the joints to increase safety in human-robot collaborative tasks.

Words/Phrases and Expressions

link 链接
carbon fibre 碳纤维
joint 接头
seal 密封
waterproof 防水
crevice 缝隙
spring-energised 弹簧驱动
plain hexagon head 普通六角头
unhygienic 不卫生的
actuator 执行机构
pneumatic cylinder 气缸
proportional 比例的
pulse width modulation 脉宽调制
solenoid valve 电磁阀
brushless DC servomotor 无刷直流伺服电机
AC servomotor 交流伺服电机
preassemble 预装
gearhead 减速机
brake 制动器
encoder 编码器

polyurethane 聚氨酯
de-coupling 解耦
planetary gearhead 行星齿轮箱
transmission 变速箱
premature 过早的
failure 失效
asymmetric 非对称
acceleration profile 加速度曲线
vice versa 反之亦然
asymmetry 不对称
trajectory 轨迹
sensor 传感器
vision system 视觉系统
smart camera 智能摄像机
discrepancy 差异
collision 碰撞
malfunction 故障
safety barrier 安全屏障
redundant 冗余
labour-intensive 劳动密集型的
endeavour 努力,竭力,尝试

Chapter 24 Automatic Control Technology in Food Industry

Food process control aids the main objectives of the food industry which focus on food safety, food quality control, increasing yield, and minimizing production cost. Various applications of advanced process control methods have significantly impacted on the food industry. The benefits derived from advanced process control technology have increasingly been appreciated. Thus, demands for process control techniques have become high in the food industry over the last decades. However, it is often difficult to automate and control food processes, partly due to the high variability of raw materials and issues with the real-time measuring and monitoring of important food process parameters and food quality characteristics. Therefore, different types of control design strategies have been developed. It is important to select the best control technique for a particular application in the food industry.

In general, reliably established relationships between raw materials, process settings, and end product outcomes are the main requirements for being able to control an industrial process. Classical control techniques are mainly represented by the proportional-integral-derivative (PID) controller which has been employed in a broad range of applications. PID controllers are applied mostly to the processes which are characterized by linear, low-order dynamics. On the other hand, advanced process control techniques target nonlinearities and high-order dynamics which are ubiquitous in many processes in the food industry. Thus, it is often necessary to employ advanced food process control for these processes. Advanced control techniques might be classified into the following three main categories: model-based control, fuzzy logic control, and artificial neural network-based control. Also, it is stated that computerized control systems and intelligent control techniques such as expert system, neuro-fuzzy control, and hybrid control system could bring advantages including easy control of complicated food processing applications and quick control actions for several food processes. Most of the process control applications in the food industry stayed relatively behind the latest theoretical developments.

In this chapter, advanced control strategies, particularly model-based controllers, fuzzy logic controllers, and neural network-based controllers are firstly described with their main characteristics. A number of applications of the advanced control strategies are then discussed according to different food processing industries such as baking, drying, fermentation/brewing, dairy, and thermal/pressure food processing.

24.1 Advanced Control Design Approaches

The aim of implementing a control system for a process is to assure the stability of the process despite of load changes and disturbances and to optimize the overall performance of the process. Advanced control techniques comprise a wide variety of methods, ranging from model-based predictive controllers to intelligent and "software" sensors, neuro-fuzzy control, and expert systems. However, it is known that traditional PID controllers (Figure 24.1) have been widely used since the 1930s due to their affordable, robust, simple, and commercially accessible advantages.

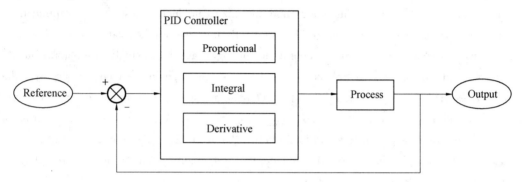

Figure 24.1 PID Control strategy illustrative block diagram

Although PID controllers have advantages, they only perform well with linear processes and low-order kinetics. Thus, those food processes that show strongly nonlinear characteristics would lead to poor behavior of this type of controllers. Therefore, advanced control strategies become more important to cope with nonlinear process dynamics.

24.2 Advanced Control Strategies

1. Adaptive Controllers

Adaptive controllers (Figure 24.2) have the capability of automatically adjusting or compensating to changes in the process or environment by regulating settings. Adaptive controllers might be classified into two types: self-tuning controllers (STCs) and model reference adaptive controllers (MRACs). For STCs, a parameter estimator is responsible for utilizing the process inputs and outputs to predict the process parameters online. A change in the process parameter is then sent to the controller to adjust its settings. MRAC is based on an error signal which is the difference between model outputs Y_m and the actual process outputs Y.

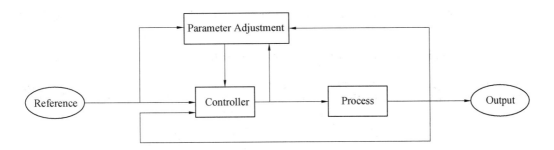

Figure 24.2 Adaptive controller illustrative block diagram

2. Intelligent Controllers

Intelligent controllers are established for those processes with higher-order dynamics. Process history, failure detection, and heuristic methods are generally employed to adjust the controller parameters. Fuzzy logic and neural networks are the most popular approaches for intelligent controllers.

3. Fuzzy Logic Controller

Fuzzy logic controller (Figure 24.3) is a rule-based approach which uses logical connectives such as IF (condition) and THEN (conclusion). Most of the biological processes follow nonlinear patterns so that fuzzy logic controller might be a better method in food processing control. Fuzzy logic controller brings a systematic methodology to convert expert knowledge into a heuristic control algorithm. Therefore, it becomes very beneficial for those food processes that are not totally understood or have limited mathematical models. In addition, complex mathematical relationships are not required in the establishing of a fuzzy logic controller.

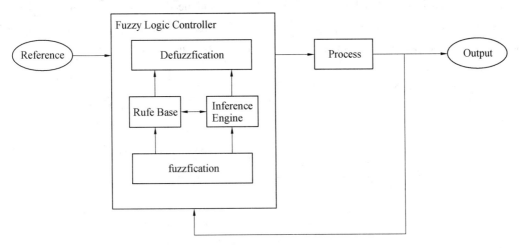

Figure 24.3 Fuzzy logic controller illustrative block diagram

4. Artificial Neural Networks

Artificial neural networks (ANNs) (Figure 24.4) are dynamic systems which are composed of a simplified version of biological neural networks in the brain. ANNs can use input-output data to recognize and guess patterns by training and learning. ANNs have unique benefits to process control including control of unknown and complex processes. However, they need substantial computational power and a relatively longer time for their initial learning period.

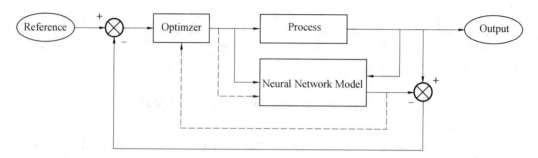

Figure 24.4 ANNs controller illustrative block diagram

5. Model-Based Control

An analytical or empirical model characterizes the attributes of a process, and according to the model, a control rule can be constructed for the process (Figure 24.5). A process model should be precise, simple, stable, and applicable for online control purposes. Due to mostly nonlinear characteristics of food processes, linear approximations may not be adequate to describe the behavior of the food processes. Therefore, nonlinear model-based controllers are advantageous where process nonlinearities are strong and require several adjustments of controller. Generic model control (GMC) was developed with the particular objective of integrating nonlinear, multivariable process models directly in the control algorithm. Analyzing the stability of nonlinear closed-loop feedback systems is a challenge and needs advanced mathematical concepts. Also, it is crucial to recognize and evaluate the effect of modeling errors on the closed-loop stability and performance when a control algorithm depends on a mathematical model.

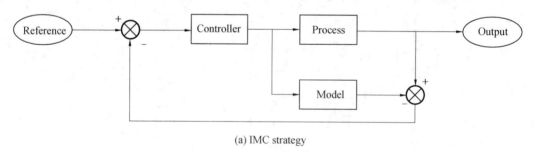

(a) IMC strategy

Figure 24.5 Model-based controller illustrative block diagram

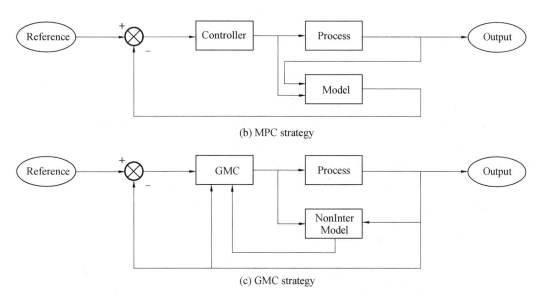

Figure 24.5 (Continued)

6. Supervisory Control and Data Acquisition

Supervisory control and data acquisition (SCADA) is a type of control system environment which collects data over a process net-work and employs control actions to various parts of the process network, including PID and some advanced controllers. In addition, SCADA can provide several other functions, e. g. , data processing, security access control, graphical user interface, communication services, trend display, and alarm management.

24.3 Applications of Advanced Control Strategies in Food Processing

24.3.1 Baking Process Control

Baking mainly involves with heat and mass transfer phenomena which result in changes in the chemical and physical properties of dough. Thus, baking is a dynamic process and the relationships between dough and baking parameters lead to various quality alterations. As a consequence, employing suitable control strategies plays a key role in delivering product quality and minimizing energy consumption during a baking process.

One of the first fuzzy control applications to a food process was on a flat bread extrusion cooking process almost three decades ago. Later, the object-oriented programming environment was implemented on a PC to create a fuzzy expert system which was used to make an extruder control system. The digital image processing could be used with an optical system for monitoring and controlling a baking process. The state of the baking process was identified reliably based on a neural network model to calculate the lightness and color saturation of bakery products.

Flat bread extrusion applications were implemented based on mechanical energy via adjusting screw speed in order to improve process yield by using fuzzy logic, neural networks, and model-based control approaches. With regards to leavened bread and biscuit baking processes, preheating steps, image processing, and operator decision support applications were used to improve product quality (e.g., lightness, color, browning, other sensory attributes) by employing neural networks and model-based control strategies. The food extrusion processes have been controlled using various approaches ranging from fuzzy logic control to model predictive control. In the baking industry, energy consumptions for the ovens are high, and cost-effective production of bakery products is highly demanded for mass manufacturing. Thus, model-based control systems are a promising approach to achieve optimum process conditions with less energy requirement to produce high-quality bakery products.

24.3.2 Drying Process Control

Drying is one of the most energy-demanding processes used in many industries including food, agriculture, and biotechnology. The main target of drying is to decrease moisture content within a product by using thermal energy to produce desired characteristics for a particular dry product. There are more than 60 000 dry products and 100 dryer types that are commonly used around the world. It is impossible to employ a single type of controller to all dryers due to the fact that different drying applications require various technical approaches.

Most of the control applications to drying targeted at higher production yield and less energy consumption for several food industry practices such as drying of grain, cereal, maize, corn, milk powder, egg, and sugar beet by controlling motor intensity, temperature, and humidity. This is expected considering the high energy demand of drying processes. With regards to product quality, drying processes have critical parameters such as moisture content, water activity, and total solids for the final product to meet tightly, which can be achieved by employing PID, fuzzy logic, ANNs, and model based controllers in the industry. Different model-based or model predictive control (MPC) strategies have been applied to grain dryers and nut roasters. More recently, latent variable model predictive control (LV–MPC) was investigated for trajectory tracking in batch processes with principal component analysis for temperature control processes which were highly energy demanding.

On the other hand, fuzzy logic and neural network based controllers have been applied to more complex applications including baker's yeast drying process and drying of cured sausage due to the fact that living microorganisms and changing product properties lead to nonlinear process dynamics or uncertainties in the process. Therefore, hybrid control systems and new learning algorithms may be more promising for the control of some drying processes.

24.3.3 Fermentation/Brewing Process Control

Fermentation processes play an important role in the food industry, and their effective production steps, monitoring, and intelligent control systems are highly demanded. Nevertheless, due to their complex process dynamics such as nonlinear attributes, presence of living microorganisms, and various end products, it is a challenge to deal with the monitoring and control of fermentation processes.

Automation has been employed in the brewing industry. For fermentation/brewing, the presence of living organisms and different final product characteristics causes many challenges in the monitoring and controlling of fermentation processes. Adaptive linear controller has attracted great interest, and it is especially suitable for fermentation and enzymatic processes in which the biocatalyst activity may change suddenly. This type of controller is based on an online estimation algorithm for automatic parameter adjustment when process conditions may vary.

Most of alcoholic fermentation applications aimed at better performance for higher yeast activity by controlling ethanol concentration, temperature, and flow rates in the processes by employing fuzzy logic, ANNs, and model-based control approaches. Control applications to manufacturing fermentation products are highly demanded due to time-dependent variations of process parameters in bioreactors. In particular, brewery applications focused on the quality of products such as consistent beer quality and foam, minimizing undesired flavor components by controlling fermentation time and temperature by using mostly fuzzy logic control and ANNs due to the high process variability sourced from raw materials. In addition, hybrid control systems including neuro-fuzzy, fuzzy-aided neural network-based control, fuzzy expert systems, and neural network-model based control have been studied for baker's yeast production.

24.3.4 Thermal and/or High-Pressure Food Processing

Most of the thermal process applications focused on lower energy requirements by controlling either heating or refrigeration systems. With regards to product quality attributes such as color, as well as product safety, thermal treatments and preserving foods at particular temperatures or pressures were the primary target of these control applications. It is of interest to note the applications of ANNs to the textural characteristics of food products by employing spectroscopic techniques and machine vision for thermal processing control applications.

Different canned food processors use automation and intelligent online control to achieve lower cost and improved safety assurance. Moreover, a neuro-fuzzy model is reported in a food thermal processing application which employed model predictive control strategies. Controllers were integrated in a pilot plant for the thermal sterilization of solid canned food in steam retorts. Recognition and classification of the various colors of food samples such as chicken wings, whole chickens, high cereal patti burgers, sausages, and minced beef burgers were investigated using a fiber optic-based spectroscopic technique coupled with artificial neural network signal processing in a large-scale industrial oven. In this application, neural networks with spectroscopic techniques were developed for each product to determine quantitatively whether a product had been cooked to the optimum color.

High-pressure processing is a growing technology and brings several advantages including the effective control of microbiological and/or enzymatic activity, better preservation of nutritional value, and negligible alterations of organoleptic properties compared to thermal processing. However, it is challenging to model the thermodynamic behavior of foods during high-pressure treatments employing first principle-based models due to the lack of thermophysical properties of foods under such pressure conditions. ANN modeling is particularly useful for systems involving complex physical-based models and unavailable thermophysical properties, e. g., those under high pressure.

Thermal and/or pressure food processing applications in terms of process control are mostly based on minimizing the consumption of energy for food manufacturing. Thermal sterilization of canned foods, tubular heat exchanger for thermal food processing, and controlling of cooling units to preserve foods at certain temperatures could be listed as typical examples for such applications. Due to the scarcity of data on food properties under high-pressure conditions, combining model-based control approaches with ANN and fuzzy logic control should particularly be useful to utilize sophisticated first principle-based models in the control of high-pressure food processing under different pressure conditions.

24.3.5 Dairy Process Control

Developments in the understanding of dairy process mechanisms and product characteristics brought in various control strategies such as model-based control and neural network-based control. Several input parameters, e. g., milk concentrate viscosity and atomization, particle size distribution, drying mechanism, and protein denaturation, were used for developing model-based control strategies in milk drying processes. It is reported that most industrial dairy manufacturing processes were working continuously at high capacity throughout the year, and thus, they required a high level of process control during the manufacturing.

Dairy control applications have focused on two main categories which are optimizing thermal pasteurization process and minimizing parameter variability such as compositional ones, by employing model-based control strategies using temperature control and sensors in the plant for the process parameters. In addition, cheese manufacturing applications targeted at ripening process by employing ANNs and fuzzy logic control strategies in order to improve sensory and quality parameters.

For dairy processes, extensively accumulated knowledge of process characteristics and product properties helps in the adoption of predictive modeling. For thermal treatments in the dairy industry, generalized predictive control and ANNs were mainly used to deal with process nonlinearity. On the other hand, the control of the cheese ripening process and decision aid systems for the cheese making process were by fuzzy logic control. Linguistic expressions helped control the processes with operator decision support. Adaptation and implementation of some advanced online measurement technologies including NIR spectroscopy and acoustic spectroscopy might induce new promising models for the predictive control of dairy processes.

24.3.6 Applications in Other Food Processing

Food processing is the core component of the modern food industry. Using different techniques, raw materials are processed to marketable food products. From raw materials to end products, generally there are several stages whose selection and connection are often unique to the products. Therefore, control strategies are required to be tailored for particular applications.

Centralized control systems are one of the main components of an automation system, and they are employed in the food processing and packaging industry. The systems are reconfigurable, which improves their abilities and functionalities the advanced field bus technology, which is a control networking system for connecting different field devices such as sensors, regulators, and controllers, allows an easy implementation of DCS and SCADA architectures.

The coffee grinding is controlled with ANNs, in which a neural network was used to control two grinders which were employed for coffee production. They concluded that in the future, this approach could be used for developing an automatic decision support system to keep coffee particle granulometry within desired levels.

24.3.7 Prospects

Advanced process control strategies become an emerging requirement for the food industry. They not only lead to improved food product quality but also bring other economic benefits such as lower-energy consumption and more cost effective production. In addition, advanced process control has an increasing role in today's food industry due to that newly developed food processing methods are often more complicated and challenging which demand for effective process control systems to deliver their benefits.

The performance of a control system is highly dependent on process models. Thus, more reliable process models are required in order to enhance the performance of advanced control systems. Advanced controllers particularly target at those food processes that have nonlinear dynamics, uncertainties, and variable time delays. Robust control of large-scale nonlinear dynamics in the food industry remains as one of the main challenges for process control applications. In addition, highly stochastic systems pose more limitations for the process control applications. Besides emerging process control applications of fuzzy logic and neural network-based controllers, modeling of food processes requires more focus to understand different dynamics in the near future.

It should also be considered that there are significant differences between laboratory-scale and industry-scale applications of different control strategies. Therefore, sensors for online control and the complexity of processes bring new challenges and opportunities to developing advanced food process control systems such as expert systems, human operators' decision systems, or hybrid systems. However, more in depth studies as well as training for effectively utilizing operator-aided systems are required to deliver the real benefits of food process control.

In order to improve the implementation of process control, one of the important efforts should be to use intelligent control systems with innovative sensors and combine them with modern methods of data analysis and modeling according to the process and product information.

24.3.8 Conclusions

In conclusion, combining different process control strategies and applying advanced computation techniques bring encouraging prospects for food processing applications. Advanced control strategies delivered promising aspects to various food processing applications with increased accuracy, maximized food product quality, and reduced production costs. In today's highly competitive industrial environment, more adoption of advanced control strategies is anticipated in the near future. Towards this, research efforts will be focused on developing various process models for the purpose of designing and implementing process controllers.

Words/Phrases and Expressions

Proportional-Integral-Derivative (PID) controller PID 控制器
linear 线性
low-order dynamic 低阶动态
process control 过程控制
nonlinearity 非线性
high-order dynamic 高阶动态
fuzzy logic control 模糊逻辑控制
Artificial Neural Network Control(ANN) 人工神经网络控制
Intelligent Control Technique 智能控制技术
expert system 专家系统
Neuro-Fuzzy Control 神经模糊控制
hybrid control 混合控制
advanced control strategy 高级控制策略
brew 酿造
adaptive controller 自适应控制器
compensate 补偿
self-tuning controllers (STCS) 自整定控制器
model reference adaptive controllers (MRACS) 模型参考自适应控制器
heuristic 启发式的
control algorithm 控制算法
generic model control (GMC) 通用模型控制
supervisory control 监控
data acquisition 数据采集

trend 趋势
alarm 报警
heat transfer 传热
mass transfer 传质
trajectory track 轨迹跟踪
in batch 批处理
NIR spectroscopy 近红外光谱
acoustic spectroscopy 声谱学
field bus 现场总线
granulometry 粒度测定，粒度分析

Chapter 25　Novel Food Packaging Technologies

Novel food packaging technologies arose as a result of consumer's desire for convenient, ready to eat, tasty and mild processed food products with extended shelf life and maintained quality. Recent trend of lifestyle changes with less time for consumers to prepare foods posed a great challenge toward food packaging sector for the evolution of novel and innovative food packaging techniques. The novel food packaging techniques, viz. active packaging, intelligent packaging and bio active packaging which involve intentional interaction with the food or its surroundings and influence on consumer's health have been the major innovations in the field of packaging technology. These novel techniques act by prolonging the shelf life, enhancing or maintaining the quality, providing indication and to regulate freshness of food product. The advancement in novel food packaging technologies involves retardation in oxidation, hindered respiratory process, prevention of microbial attack, prevention of moisture infusion, use of CO_2 scavengers/emitters, ethylene scavengers, aroma emitters, time-temperature sensors, ripeness indicators, biosensors and sustained release of antioxidants during storage. The novel food packaging technologies besides the basic function of containment increase the margin of food quality and safety. The novel food packaging techniques thus help in fulfilling the demands throughout the food supply chain by gearing up toward persons own lifestyle.

25.1　Active Packaging

Active packaging came into existence with the aim of satisfying the consumer demand for natural, recyclable, and biodegradable packaging materials. Thus renewable resource based active packaging material capable of degrading by natural compositing process and with less environmental effect was developed. Active packaging prolongs the storage life and enhances the margin of food safety by altering the condition of the food. Active packaging is used as a substitute to conventional food processing techniques (high thermal treatments, brining, acidification, dehydration and additive preservation). The basic underlying principle behind the use of active packaging depends on the incorporation of particular components inside the polymer and intrinsic characteristics of the polymer itself used as packaging vehicle.

A new advancement in the use of active packaging is the addition of polymer materials containing some additives that impart anti microbial properties. These polymeric matrices have the potential of releasing active agents (antioxidants and antimicrobials), retaining compounds (ethylene, oxygen and water) or undesirable food components. The potential scavengers like cyclodextrins used in the latter application act irreversibly and are either inorganic metals or salts. Controlled delivery of active agents into the food via packaging films for extended periods of storage and distribution restricts the development of undesirable flavors produced as a result of directly incorporating additives into the food. The use of artificial antioxidative agents like butylated hydroxy toluene, thioester and organophosphate compounds as active packaging additives is limited due to toxicity as a result of their migration into the food products. Hence, the use of synthetic additives is now replaced by the use of essential oils as natural extracts obtained from herbs and spices, tocopherol and extracts from plant which are generally recognized as safe and enable the chemical stability of oxygen sensitive foods. In active packaging system choosing antioxidant, antimicrobial, oxygen scavenger and moisture absorbing agent is critical which requires full considerations.

25.2 Intelligent Packaging

Intelligent packaging is rooted on involvement of intentional association of food with its package or surroundings with an attempt to enhance food quality characteristics and safety. Intelligent packaging is linked to the advancement in time—temperature regulators, ripeness monitors, biosensors and radio frequency indicators and regulators. Therefore, intelligent packaging provides signal for perceiving and evaluating the freshness of food.

Intelligent packaging provides knowledge about properties of the enclosed food and its existing environment and helps in providing basic idea to the retailer, customer and manufacturer about the state of these properties. Intelligent packaging provides additional function to the basic communication function of conventional packaging because it provides knowledge to the consumer on the basis of its ability to observe or record internal and external changes in the product surroundings.

The two important functions performed by intelligent packaging are to monitor both internal and external conditions that is to record changes occurring both outside and inside the packaging. The latter function of intelligent packaging that is assessing the quality of the food product directly within package involves intimate association with the headspace or food which necessitates the use of indicators for the safety and quality of packaged food item. Typical indicators represent signaling gas leakage, ripeness regulators and indicators, time-temperature monitors, bio probes, radio frequency indicators and toxin indicators. The concept of intelligent packaging in real sense is to evaluateefficacy and strength of active packaging system. Intelligent packaging offers greater significance by providing detailed knowledge throughout the supply chain and maintains food quality by finding out critical points by the use of attached, incorporated or printed labels onto packaging material. For example, the temperature regulators in self heating and cooling systems, and ethylene adsorbers and absorbers that can be supplied either as sachets or added to the package itself.

Radio frequency identification (RFID) used in intelligent packaging is radio wave based system that wirelessly tracks items. RFID consists of readers (receivers), labels or radar (data carriers) and computer software, hardware, net working, and database, like the barcodeor QR code system. Use of RFID in food industry has emerged as a gateway, starting from monitoring of food to its traceability in order to enhance food safety and improve supply chain effectiveness. In the food industry accelerated rate and effectiveness of RFID technology in stock rotation and traceability of several commodities throughout the supply chain led to increased on-shelf availability at the retail level.

Besides these important forms intelligent packaging has taken role in all fields of nanotechnology and resulted in quality and safety monitoring along with sustainability of packaging.

25.3 Bioactive Packaging

Bioactive packaging is the novel packaging technology that alters the package or coating in a way so as to have positive effect on consumer's health. Various techniques known to retain characteristic properties of biopolymers and employed in this novel packaging approach include enzyme encapsulation, nanoencapsulation, microencapsulation and enzyme immobilization. Keeping in view the required functional properties of particular bioactive components, functional or bioactive packaging has the potential to maintain bioactive substances in desired proportions until their controlled or fast diffusion within the packed food during its storage or prior to its consumption. Process of bioactive packaging technology is implemented via:

(1) Utilization of biodegradable packaging materials for the release of functional or bioactive components;

(2) Encapsulating bio active ingredients into the foods or to the packaging materials;

(3) Introducing packaging materials exhibiting enzyme activity and capable of transforming some food components in order to deliver health benefits.

The development of such packaging systems exerting health promotion effect involves the concept of marine oils, prebiotics, probiotics, encapsulated vitamins, phytochemicals, lactose free foods, bioavailable flavonoids and many more will boost the packaging industry in near future because of growing human health consciousness.

25.4 Innovative Packaging Technologies

25.4.1 Functional Barrier

Functional barrier consists of one or more layers of food-contact materials and it should ensure the compliance of product with regulation. As, per definition the substances at the rear side of functional barrier will not, migrate in the food and thus will not have deleterious effects on human health nor will result in unacceptable changes in the composition and organoleptic properties. In case of articles for infants and other susceptible persons, the prescribed limit of unauthorized substances that might through the functional barrier should not exceed 0.01 mg/kg food.

25.4.2 High Chemical Barrier Material Innovations

The quality of food can be maintained by preventing adsorption, desorption, diffusion of gases, liquids, penetration of other molecules such as oxygen, pressurized liquid or gas, and water vapor by the use of high-barrier packaging. The process of polymer blending, coating, lamination, or metallization is used to enhance the barrier property of packaging materials by combining the package materials with other high-barrier materials. The structural network of blend of packaging material with high barrier packaging material affects its permeability. An example of used blends is aluminum-metallization on PET. The innovative technique include epoxy spray on PET bottles, transparent vacuum-deposited or plasma-deposited coating of silica oxide on PET films and composites of plastics with nanoparticles.

25.4.3 Intelligent Supply Chain

In developing newer value added services, supply chain provides a provision of increasingefficiency by automating simple and valuable data flows. This intelligent supply chain can lay down flat form for value addition of fresh products. In response to larger retailer mandates and compliance with regulatory bodies requirements a Spanish company ECOMOVISTAND developed an innovative and ecological packaging and transport unit, called MT, for the grocery supply chain, which can be used in the entire product cycle. That is, the MT serves: as packaging at the producer; as transport unit; as storage at warehouses; as display stand at the supermarket, all in the same mechanical system, being thus a Returnable Packaging and Transport Unit.

25.5 Interactions of Active/Intelligent Packaging with Supply Chain

A special feature of supply chain is inclusion of several actors together for sound collaboration, coordination, and information exchanges between them for better efficiency and productivity. The major problem faced in transportation of boxes, containers, pallets and cases is lack of information and control on their status influenced by the actors in the supply chain. Some large container and pallet producing companies encounter economical and logistics problems to provide on time service due to lack of information on where a pallet is and for how long it has been there. Thus, it does not seem astonishing why major retailers put thrust to come up with this lack of regulation and control by pushing suppliers toward the implementation of newer appropriate technologies.

25.6 Nanotechnologies in Food Packaging

Nanotechnology has proven most promising innovative technique by introducing latest enhancements in food packaging by providing mechanical and barrier properties, detecting pathogens and introducing smart and active packaging keeping in consideration food quality and safety aspects. Presently, the nanotechnology that is playing part in the market is the nanolayer of aluminum that coats the interior of many snack food packages.

Among the various nanomaterials the most promising for food packaging is nanocomposites. Nanocomposite packages for food have taken their place in the market and many are yet to be launched to contribute major portion in the future to food packaging. Nanocomposite materials have played a vital role in improving the strength, barrier properties, antimicrobial properties, and stability to heat and cold (fundamental properties) of food packaging materials. For other Nanotechnologies, the mechanical strength of food packaging materials can be improved by incorporation of carbon nanotubes of diameter in nanometers which are cylindrical in shape with antimicrobial properties. It was found that these carbon nanotubes resulted in cellular damage in Escherichia coli by puncturing cells and eventually leading to their death.

25.7 Food Safety Issues

The present food legislation keeping in consideration the consumer desire for natural, minimally processed and convenient food products in addition to un ending changes at industrial, retail and distribution levels considers food safety as a global concern. This concern provides a stimulus to the packaging industry in order to present numerous innovative techniques to tackle with the legal and regulatory requirements along with the changing needs of the food industry and consumers.

In food supply chain the concern of food safety and quality has given way to exercise more control and explore information within supply chains as well as to the consumers on processing, sourcing and distribution of food products.

In logistic chain for chilled food, involving on board handling in ships, air transport, an attempt is that several companies use instruments, such as RFID tags with embedded temperature sensors physical and chemical sensors to monitor the temperature for traceability information.

The safety concern related to active and intelligent packaging should be addressed based on following three important considerations.

(1) Labeling, proper labeling should be done in order to prevent misuse and misunderstanding by the consumers or downstream users, e.g. to prevent sachets from being eaten.

(2) The migration of active and intelligent substances with respect to their toxicity should

be kept in consideration and their migration process should comply with food legislation. Monitoring the phenomena of migration means to adapt some mass transfer modeling tools and migration tests other than those applied or recommended for conventional plastics, as they cannot be adapted to active and intelligent systems.

(3) Efficient packaging, most importantly the claimed function of food packaging in few cases can give rise to safety concern as for any food preservation technology. Delivering a preservative or absorbing oxygen in a suitable way for preventing microbial growth without inducing antimicrobial resistance or pathogen over growth.

25.8 Environmental Issues (Biosourced, Biodegradable, Recyclable)

The environmental policy objectives include decrease or even to prevent the use of packaging, to recover and recycle all residues, and to make the producer responsible for the waste, as well as for the costs of recovering and recycling. These environmental policies certainly add extra cost all over the supply chain but are equally important for a sustainable growth. The use of returnable transport units in addition to operational and ecological benefits will help to comply with waste regulation.

There is an increasing demand for identifying biodegradable packaging materials and finding innovative methods to make plastic degradable. Biodegradation is the process by which carbon-containing chemical compounds are decomposed in the presence of enzymes secreted by living organisms. The use of bioplastic is to replicate the life cycle of biomass by conserving the fossil fuels, carbon dioxide and water production. There are three requirements for the fast degradation process viz. temperature, humidity and type of microbes. It is expected that global market for biodegradable polymers would rise at an average annual growth rate. Starch, cellulose, poly-beta-hydroxyalkanoates (PHB) and polylactide acid (PLA) plastics are the acceptable biodegradable materials used in food packages.

25.9 Future Trends

Nanotechnology is likely to play important part in the near future keeping in consideration the safety concern associated with packaging. To address the safety as well as other additional issues research and development in the field of active and intelligent packaging grew at dynamic pace with the aim to provide ecofriendly packaging alternatives. This posed a challenge of designing packaging materials by employing reverse engineering approach on the basis of requirements of food product besides on the availability of packaging materials. The aforementioned approach resulted in the tailoring of stimulated/controlled release of active agents and for specific target indicators. Another area of development is the use of innovative non-migratory materials in case of functional in-package food processing.

25.9.1 Future Advances of Active Packaging

The advancement in the area of active food packaging led to the development of stimuli-responsive polymer materials. These unique materials offer amazing, innovative and functional features that fully comply with existing environments and regulate the release of molecules in response to external stimuli. These stimuli-responsive macromolecular nanostructures are tailored to bring about conformational as well as chemical changes as a reaction to external stimuli such as change in chemical composition, temperature or pH.

25.9.2 Edible Coatings

Edible films and coatings offer huge future potential to satisfy the consumer desire for environment friendly and natural foods. They do not completely replace traditional food pack aging materials but provide extra functionalities to the food. Since, these packaging materials are produced from agricultural wastes and/or commodities of industrial food production, thus impart value addition to biomass. Use of edible films and coatings can enhance the process of preservation of food in addition to reducing the traditional packaging both in cost and bulk.

Edible coatings and films are developed from biopolymer based on hydrocolloids, such as polysaccharides like cellulose, starch, alginates, chitosan, gums, pectins and proteins, from vegetable or animal origin. In addition to the basicfunctional properties of providing barrier to gases and moisture the new innovative development includes use of composites or blends to regulate the release of food additives and nutrients. These edible films can be enriched with additional functional features, including regulating the release of food additives and nutrients, such as properties of anti-oil obsorption, anti-fat obsorption and anti-microbial.

Words/Phrases and Expressions

retardation 迟滞,迟缓
hinder 阻碍
respiratory 呼吸的
scavenger 清除剂
ethylene scavenger 乙烯清除剂
emitter 发射器
aroma emitter 香气发射器
ripeness indicator 成熟度指示器
antioxidant 抗氧化剂
active packaging 主动包装
recyclable 可回收的
biodegradable 可生物降解的
brine 盐渍

acidification 酸化
dehydrate 脱水
additive 添加剂
preservation 保存
antimicrobial 抗菌剂
cyclodextrin 环糊精
inorganic metal 无机金属
butylated hydroxy toluene 丁基羟基甲苯
thioester 硫酯
organophosphate 有机磷酸盐
toxicity 毒性
essential oil 精油
herb 香草
spice 香料
tocopherol 生育酚
bio probe 生物探针
barcode 条形码
DR code 二维码
traceability 可追溯性
bioactive packaging 生物活性包装
enzyme encapsulation 酶包封
nanoencapsulation 纳米包封
microencapsulation 微包封
enzyme immobilization 酶固化
enzyme activity 酶的活性
marine oil 海洋油
prebiotic 益生元
probiotic 益生菌
encapsulated vitamin 封装的维生素
phytochemical 植物化学物质
lactose free food 无乳糖食品
bioavailable 可生物利用的
flavonoid 类黄酮
barrier property 阻隔性能
detect pathogen 检测病原体
smart and active package 智能主动包装
nanocomposite material 纳米复合材料
carbon nanotube 碳纳米管

cellular damage 细胞损伤
escherichia coli 大肠杆菌
puncture 穿刺
supply chain 供应链
logistic chain 物流链
chilled food 冷藏食品
embedded 嵌入式的
biosourced 生物来源
residue 残留物
decompose 分解
fossil fuel 化石燃料
cellulose 纤维素
poly-beta-hydroxyalkanoates（PHB）聚 β-羟基链烷酸酯（PHB）
polylactide acid（PLA）聚乳酸（PLA）
edible coating 食用涂料
edible film 食用薄膜
biopolymer 生物聚合物
hydrocolloid 水胶体
polysaccharide 多糖
alginate 海藻酸盐
chitosan 壳聚糖
gum 胶质物
pectin 果胶
anti-oil obsorption 抗吸油
anti-fat obsorption 抗脂吸收
anti-microbial 抗菌

Chapter 26 Evolution of the Food Industry—People, Tools and Machines

26.1 Introduction

The past, present and future in the food industry-always an evolving relationship between people, tools and machines to prepare food. Technical progress has seen a development from first simple tools, which were used about 770 000 years ago to cook meat, to current day procedures with often automated equipment. The progress, especially seen in the late 19th and the early 20th century, was driven by the increasing demands of consumers for both attractive, nutritious and flavor some food.

This change from primitive tools to sophisticated machines has met demands for better quality of products that were to become available in many corners of the world. But as the 20th century progressed, such developments also brought with them other factors that must be mastered and kept under control. The understanding of what contaminants lead to food safety problems were added to the requirements that had to be built into tools and machines used in the food industry. Since the mid 20th century, progress in design focused as a first priority on the non-negotiable: the ability to produce safe food.

26.2 Evolving Relationships Between People and Their Tools and Machines

When tools were mainly manual, they were rudimentary and designed for the one purpose of handling and preparing food. During the industrial revolutions, man-made machines appeared. These machines were valued and each user sensed a responsibility and pride in maintaining and cleaning them. Even though machines became more mechanical and more independent of people "to turn the wheel", they were still regarded as valuable. They were difficult to replace and were meant to last years. Many centuries later many of these early machines still continue to function well.

But as processes became more automated, the interdependence of machines and people was replaced by a certain independence of equipment from operators. This caused a perceived possible relaxation of the relationship between people and machines. Maybe this led people to show less care, less pride and less value for machines, which were now "doing the job alone". An example was the development of the "on-off push-button complacency" in relation to CIP of equipment.

However, the food industry could just not operate this way. Equipment in a food factory is simply not a robot. From its conception as well as during operation, equipment must be ensured to be safe and to produce safe products. A point was reached where people realised that again more attention had to be paid to machines which are in contact with food and to have control over its quality. Not only design but also installation, cleanability and maintenance were seen as critical for ensuring safe and consistently nutritious and flavorsome food.

Tools and machines cannot produce quality on their own. As in the first days of food handling for consumption, the interdependence of tools and machines with the handler must still exist. People must take charge to assure safe and consistent quality is produced.

Some companies and EHEDG have been promoting such concepts for several years. But it is important that the entire food industry adopts necessary measures of prevention. The starting point is to ensure that the tools and machines and the industrial services supplying important elements for the process are engineered directly in line with the needs of the final product and its consumer.

26.3 Machine Selection: Focus on the Product

To write down all that is needed to appropriately engineer, operate and maintain tools and machines requires many manuals. So this will be a more general account of where to focus efforts. These facts are based on many years of experience of prevention of problems caused by inadequate hygienic engineering and in many cases, by a lack of understanding of what really matters for food safety and quality. To simplify the steps to take, it is essential to "Focus on the product and all that is in contact with the product".

When reference is made to focus on the product, the illusion is created in some cases that all this is taken into consideration in HACCP. This can be true but often people fall into the trap of concentrating on the CCPs and forgetting the prerequisites, many of which are related to hygienic engineering of product contact surfaces. As with HACCP, the list of what must be considered for hygienic engineering to ensure food is safe and of consistent quality begins with the product and consumer type.

Product type: process details and its consumer.

Product contact materials of the tools and machines.

Product contact services: air, water.

Product contact surface cleaning.

Product contact surface maintenance: routines, materials, e.g., lubrication.

All these must be under control but this cannot be achieved without the last and important point: the people factor.

26.4 Product Type: Process Details and Consumers

1. Product Type

Certain equipment are specifically designed for one type of product such as those found for moulding chocolate or ice cream extrusion. But the majority of equipment can be used for a wide range of products and target consumers, from infants to adults. This implies that either the equipment has to be designed for the most exacting of the requirements or, in practical terms and in relation to investment, to have additional features to the basic design to address predicted higher risks.

Take the Example of a UHT Filling Equipment.

This could be used for acid fruit juices, chocolate milk and even infant formulae. For the former, food safety risks are minor and equipment design should meet the basic requirements of good engineering. On the other hand, chocolate milk, due to its characteristics, makes a machine much more difficult to clean. It requires added features related to cleanability, with special attention to product contact seals and valves, possibly requiring some modifications. For infant formulae, where no risks can be permitted, all factors that could contribute to deviation from strict food safety requirements must be under control.

Taking these two latter examples, the equipment supplier cannot predict for which purpose a machine may be used. There has to be a very close relationship established between supplier and client in the product development stage so that features necessary for a particular product's food safety are built in. This may seem a logical step to take but how often have factories received ordered equipment and later found incompatibility problems between equipment and the product?

2. Wet or Dry Process

The choice of a particular type of equipment also depends on its planned use, either a wet or an essentially dry process. For a dry product, it is well known that a critical preventive measure for re-contamination is to limit wet cleaning and to only dry clean. The latter requires a different set of equipment specifications, particularly because accessibility for dry cleaning brings with it a totally different set of design parameters compared to those required for CIP or manual wet cleaning.

Such points must be defined in the specifications to the supplier and brought out in discussions before actual purchase is finalised.

The key word to success is transparency. Without elucidating confidential details of a new process, information exchange up front should permit acquisition of an equipment of suitable hygienic design for the process in question.

26.5 Product Contact Materials

In an ideal world, all equipment would be made of stainless steel but certain technologies require different materials for product contact surfaces. It is quite common to still find high chromium content materials in ice cream freezers, extruders and surfaces of roller dryers.

In this case, design and cleanability/cleaning routines have to be closely linked. If not and the equipment is connected, for example, to a common central CIP using acid and alkali or worse still, chlorinated alkali, serious pitting of surfaces can occur. This cannot be attributed entirely to faulty design or installation but to the lack of understanding that one solution for cleaning is not appropriate.

A certain responsibility does rest with the supplier. They should know that some fabrication materials of their equipment have a susceptibility to certain chemicals and they should warn clients and advise them on the appropriate agents to use. Unfortunately, cleanability testing of equipment still lacks a certain priority in some suppliers' validation lists.

26.6 Product Contact: Air and Water

Historically, classic food safety cases have illustrated lessons on prevention, citing the cause as contaminants carried by process air and water. Such experience must not be forgotten. In analysing food safety preventive measures for a process, required quality of water and air must always be reviewed.

Perfectly designed equipment may still produce contaminated product if cooling or chocolate tempering water carry pathogens or the cooling air is too humid and carries salmonella. Some examples are the following:

Plate heat exchangers are generally designed and set up to operate such that no cross-contamination can occur from the cooling water to product. However, just relying on design and product over pressure is risky. Therefore, although having only indirect product contact risks, cooling water must be of a quality that is in line with product specifications. This is the only way to make certain this process step is under control.

Note: Simply specifying potable water can still lead to problems. Potable water specifications tend to concentrate on pathogens. However, if the cooling water is for a UHT product, contamination with spores can lead to important product spoilage.

Another case of indirect contact is that of tempering water for chocolate. There should be no product contact but this can occur. In the 1970s and 1980s, some widespread salmonella infections caused by chocolate bars were traced to a contaminant coming from tempering water. In another past case, the corrosion bacteria first made tiny holes between the tempering jacket and the chocolate which then allowed Streptococcus faecalis in the water access to the chocolate.

Therefore, not only should the water be of potable quality, but also it should not have any corrosion bacteria.

Cooling air for powder has direct contact with the product. As such air is often about 25 ℃ when entering a warm building, it can condense in ducts, if the RH% (relative humidity) has not been correctly adjusted. If in addition to the condensation, filters do not eliminate dust which may contain pathogens, growth may occur, resulting in air carrying unacceptable bacteria onto the powders.

These cases illustrate the importance of going back to the quality of the source of the service element and the various factors that must be controlled and maintained to ensure product does not become contaminated. The manufacturer of a plate heat exchanger or a water-jacketed tank must ensure that the finish will not lead to leaks. A supplier of powder drying or powder transport equipment may give advice on appropriate air quality. But in both cases, the user cannot just blindly believe that all will be under control when the equipment is installed and running. He has to make certain that the services are maintained in an appropriate state.

26.7 Product Contact Surface Cleaning

This is often a forgotten element during qualification of design and installation. Reference is made once again to the attitude of "on-off push-button complacency" due to installations "doing the job alone", as with CIP of equipment. Many companies can probably cite cases where such a situation has led to a product contamination. Just a simple change in the line—new pump, valve, monitoring device or sampling tap—can mean that a once robust CIP system can be adversely affected with the undesirable result of contaminated products.

It would be ideal if all equipment came with instructions for cleaning. However, this is an impossible task for suppliers, especially where equipment can be used for a variety of process lines and for diverse products. Cleaning is very much linked to type of residue to be removed.

However, suppliers can make certain that the basic needs for cleanability are satisfied through some simple standard tests (see EHEDG testing).

The client then has to take this to the next level of industrialisation in the factory with the choice of appropriate chemicals and the creation of an SOP (Standard Operating Procedure) for the type of residue to be removed.

Once installation and SOP are established, validation is essential. Such validation must be repeated every time a change, albeit small, is made to the line.

26.8 Product Contact Surface Maintenance

All installations require maintenance. One of the major sources of consumer complaints is foreign bodies. It is possible to put metal detectors and X-rays in line. But these are the last net, not the solution. Interestingly enough, even though stainless steel is recommended for equipment, this material has one of the lower detection records. Easiest to detect is gold but no equipment would be made of such material! Mild steel, on the other hand, notorious for other characteristics such as corrosion, is more detectable than stainless steel.

Most notorious are the screw feeds. These may be well engineered and smoothly finished but during operation possible problems can come to the surface. For example, a slight off-alignment can quickly lead to metal pieces being scraped off into product.

The need for maintenance and prevention of foreign bodies really highlights the point raised about the necessary interdependence of machines and people. Before adding any metal detector to a line, the first preventive step is for people to study in detail, to look and listen for surface wear at every machine and installation with product contact points.

The user must be aware and make note of corrections to prevent recurrence of the problem. Feedback to suppliers could help develop a system to prevent off-alignment. Users have a proactive role in supplying information about chronic faults, which can only help the whole industry in keeping such problems under control.

Another part of maintenance is lubrication. Most are aware of the need to use food grade lubricants at points with potential product contact. However, the rule should be: right lubricants in the right places, whilst remembering that although food grade, lubricants are not part of the recipe.

26.9 The People Factor

Especially in the cleaning and maintenance aspects, the people factor plays an important role. This is not related to personnel hygiene, which is part of any manufacturer's basic routines and GMP. In this article, the emphasis is on the personnel's care and responsibility for the hygienic status of machines and tools. There is no place for "push-button complacency" when ensuring safety of food, however automated the process and the cleaning may be.

Banishing such an attitude requires that all operators understand where and what they have to monitor and to intervene or tell someone responsible about suspect deviations.

26.10 Conclusions

Returning to the historical account of evolving people and tools, there is a need to reawaken understanding of the importance of the people factor and its impact on prevention. Pride in one's work and in one's work tools is essential. Training on hygienic engineering of all concerned, suppliers and users, is critical.

One need not waste time looking back to decide what is necessary to build for the future. Preventive measures that exist today are built on those that were successfully used in the past to prevent re-occurrence of certain contamination problems.

As for the people factor, here is a quote from the General rules for Nestle's first factory: "Each operator must monitor, check, clean and grease with care the part of the machine for which he/she is responsible and when something does not work well, he/she must make his/her boss immediately aware of the problem. In case of negligence to alert someone to the problem, the repair will be made at the cost of the operator!" (Already then talking of accountability.)

This is where coaching of operators, the final users of the tools and machines, plays such an important role. By applying hygienic engineering principles, a perfect design can be built into a process line. However, if the operator does not understand the important measures to take—the how and why—then products will still be contaminated however performant and automated a machine may be.

Words/Phrases and Expressions

filling equipment 灌装设备
infant formulae 婴儿配方
valve 阀门
incompatibility 不兼容
transparency 透明度
acid and alkali 酸和碱
chlorinated alkali 氯化碱
salmonella 沙门氏菌
spoilage 变质
streptococcus faecalis 粪链球菌
standard operating procedure 标准操作程序
albeit 尽管
metal detector 金属探测器
notorious 臭名昭著的
slight off-alignment 轻微偏移
negligence 疏忽

Chapter 27 New-Generation Digitalization in Food Industry—Case Study

In recent years, manufacturing companies have been faced various challenges related to volatile demand and changing requirements from customer as well as suppliers. This trend has a direct impact on the value chain. New technological roadmaps and suggested interventions in manufacturing systems are implemented to overcome these challenges, such as the German high-tech strategy "Industry 4.0".

Industry 4.0 principles within the food industry aim at exploiting the high innovation and economic potential resulting from the continuing impact of rapidly advancing information and communication technology (ICT) in industry. It presents many types of challenges and opportunities, an example is the introduction and integration of new technologies in order to improve quality, efficiency and competitiveness. The overall goal toward digitalization, with a particular focus on the design and manufacturing processes will help food companies face global challenges calmly, which can be met with the support of the information technologies (IT).

1. New Digitalization in Food Enterprises

(1) Manufacturing Value Modelling Methodology (MVMM) to evaluate the company and market state.

Manufacturing Value Modelling Methodology (MVMM) has been utilized as a basic tool to evaluate the current state of a food company with respect to digitalization process. The case study concerns one of the main Italian food manufacturers. Its products are produced, distributed and sold around the world and it is the Italian leader in this sector. It is an international group with sales in more than 100 countries. The company has 42 production sites, 14 in Italy and 28 abroad, which produce more than 1 800 000 tons of food products every year. A world leader in pasta and ready to use sauces in continental Europe, bakery products in Italy and crispbread in Scandinavia, the products are recognized worldwide as firm family favorites. The company has various industrial plants placed in different countries. In this case study we looked into a production plant situated in Italy. This specific company is the leader in the Italian market; it has a large global market share. With the digitalization, the food company seeks continuous improvements and has an excellent record in safety and quality of the products.

(2) The digitisation of food manufacturing to reduce waste—a ready meal factory.

A case study of a food factory producing ready meals has been utilised to demonstrate the benefits of the proposed Digital Food Waste Monitoring and Tracking System. This factory based near London, United Kingdom produces between 100 000 and 120 000 chilled ready-meals per day of various cuisines over two shifts. This factory produces approxmately 12 000 Chicken Tikka Masala (350 gm) ready meals on an average per day. Regardless of the factory strategy to reduce FW(food waste), this particular line produced around 1 400 kg of production waste per week. It was a significant problem for the factory due to the financial costs and environmental impacts of FW generated from this particular line.

①Monitoring and recording FW system. The waste was generated at various locations within the factory. Approximately 24% of FW was created in the factory due to overproduction or overstocking which resulted in ingredients going out of acceptable life and spoilage. The other restrictions such as allergen changeover, the labour movement, workload, standard batch sizes and order commitments made to ingredient suppliers added to the challenges.

FW monitoring was based on digitised "Food Waste Tracker" system. The belief was that "Food Waste Tracker" system information would be an important tool to categorise the various waste hotspots within the factory and the financial losses incurred due to FW.

Finally, production FW was minimised by 50% through daily FW tracking, staff training, and engagement and adopting the implementation of better waste solutions.

②The tracking process. FW Tracker system was placed in the production area, and all the FW related to Chicken Tikka Masala was recorded from all the departments. FW weights were displayed on the touch-screen and staff used it to record the type of waste and reason for its disposal. The software application automatically recorded the date, time and calculated its financial value in the background. The process of recording FW required less than 4.5 min per employee per week. The factory management did not have to employ any dedicated personnel to measure FW, and this practice of tracking may have reduced overall labour by minimising waste and overproduction.

2. Intelligent Packaging

(1) Deep Learning(DL)—based approach for forecasting fresh products packaging.

Fresh product packaging is a highly demanding sector, where production performance forecasting is pivotal. It is a very demanding scenario where high throughput is mandatory; on the other hand, due to strict hygiene requirements, unexpected down time caused by packaging machines can lead to huge product waste. In this context, Predictive Maintenance (PdM) technologies can be extremely valuable. PdM is a particularly relevant technology for cost reduction and production optimization in the Industry 4.0 scenario. Indeed, PdM solutions provide estimations of the health status of a machine, an information that can be exploited for taking optimized maintenance/service actions and production/logistics-related decisions. Generally, PdM solutions are developed with the usage of Machine Learning approaches and by exploiting the availability of historical data. For equipment providers, whose service-related challenges are complicated by the logistics associated with the geographical distribution of the installed machines, PdM is even more relevant.

In a real industrial case of an automatic filling machine, pieces of packaging equipment are monitored in real-time by means of an IoT framework that acquires information about current production stage, such as process settings, warnings and alarms generated by the equipment. Recently, the industrial partner has developed acloud-based infrastructure. Through a large amount of measurement data, including time intervals, operating status, alarm types, which are monitored in real-time, Deep Learning-based forecasting was employed. The aim is predicting future values of key performance indexes such as Machine Mechanical Efficiency (MME) and Overall Equipment Effectiveness (OEE).

It shows advantages with respect to current policies implemented by the industrial partner both in terms of forecasting accuracy and maintenance costs. The proposed architecture is shown to be effective on a real case study and it enables the development of predictive services in the area of Predictive Maintenance and Quality Monitoring for packaging equipment providers.

(2) Intelligent packaging for poultry industry.

In an earlier report, it was projected that the world poultry production would increase 121% from 2005 to 2050. The United States places number one in poultry meat production worldwide. Americans consume more chicken every year than any other meat. It has been reported that broiler consumption per capita has increased from 23.6 pounds in 1960 to 92.4 pounds in 2018. This tremendous success of the US poultry industry can be attributed to organizational transformation to vertical integration, technological improvement in production to processing, and continuing responsiveness to consumer demands.

Notably, when poultry or poultry product leaves the processing plant, it travels with the package to reach the consumer. The novel and innovative technologies in appropriate packaging is highly desirable.

The most technologically advanced invention in the packaging industry is the intelligent packaging (IP), in which sensors, indicators, electronic labels, tags, codes, etc. are incorporated into the packaging materials. Some examples of IP include time-temperature indicators, biosensors, gas sensors, and data carriers. These IP materials can sense or monitor the internal or external environment such as temperature, integrity, and freshness of packaged foods, record changes, and then convey the information to the processors, retailers and/or consumers.

Thus, using IP technology in the poultry meat and meat products has the potential to Maximize Poultry Product Food Safety, Monitoring Quality and Shelf life, Tracking Throughout Supply Chain and Minimizing Recall, Maximizing Consumer Engagement, and Expanding International Market. It has been reported that IP generated 23.5 billion USD in revenues in 2015 and is projected to grow 11% annually, totaling 39.7 billion USD by 2020. Collectively, these improvements will greatly benefit the industry by increasing product sale, building consumer trust, and expanding market size nationally and internationally.

References

[1] RICHARD B, KEITH N. Mechanical engineering design [M]. 10th ed. New York: McGraw-Hill, 2015.

[2] GEORGE D, LINDA S. Engineering design [M]. 5th ed. New York: McGraw-Hill, 2012.

[3] ROBERT N. Design of machinery [M]. 4th ed. New York: McGraw-Hill, 2007.

[4] TEDRIC H, MICHEAL K. Rolling bearing analysis [M]. 5th ed. Boca Raton: CRC Press, 2006.

[5] PETER H, ERIC H, BRIAN J, et al. Precision machining technology [M]. New York: Delmar Cengage Learning, 2012.

[6] CLAIRE S. Process engineering equipment handbook [M]. New York: McGraw-Hill, 2001.

[7] PAHL G, BEITZ W, FELDHUSEN J, et al. Engineering design [M]. 3rd ed. London: Springer-Verlag, 2007.

[8] FRITZ K. Manufacturing processes [M]. Berlin Heidelberg: Springer, 2011.

[9] DAVIM P. Modern manufacturing engineering [M]. Basel: Springer, 2015.

[10] ERNIE C. The lathe book [M]. Newtown: The Taunton Press, 2001.

[11] DAVID U. The mechanical design process [M]. 4th ed. New York: McGraw-Hill, 2010.

[12] SEROPE K, STEVEN S. Manufacturing engineering and technology [M]. 5th ed. Upper Saddle River: Pearson Education, Inc, 2006.

[13] SHERIF E. Processes and design for manufacturing [M]. 3rd ed. Boca Raton: CRC Press, 2019.

[14] YORAM K. The Global manufacturing revolution [M]. Hoboken: John Wiley & Sons, Inc, 2010.

[15] PAUL W. 21st century manufacturing [M]. Upper Saddle River: Prentice Hall, 2002.

[16] ROBERT D, THOMAS E. Manufacturing Technology Today and Tomorrow [M]. New York: McGraw-Hill, 1991.

[17] MYER K. Handbook of farm, dairy, and food machinery [M]. New York: William Andrew, 2007.

[18] DENNIS H, RICHARD H. Principles of food processing [M]. New York: Chapman & Hall, 1997.

[19] PAUL S, DENNIS H. Introduction to food engineering [M]. 3rd ed. London: Academic Press. Inc, 2001.

[20] ROBERT B. The role of robots in the food industry: a review [J]. Industrial Robot: An International Journal, 2009, 36(6): 531-536.

[21] YOSHIHIRO K. Robots at FOOMA Japan: a food machinery and technology exhibition [J]. Industrial Robot: An International Journal, 2011, 38(6): 572-576.

[22] MAJID I, NAYIK G, DAR S, et al. Novel food packaging technologies: Innovations and future prospective[J]. Journal of the Saudi Society of Agricultural Sciences, 2018, 17(4): 454-462.

[23] TURKAY K, WEIBIAO Z. Recent applications of advanced control techniques in food industry[J]. Food and Bioprocess Technology, 2016, 10(3): 522-542.

[24] MAGGIE R. Evolution of the food industry—people, tools and machines[J]. Trends in Food Science & Technology, 2007, 18: S9-S12.